LOST
COAL DISTRICT
OF
GEBO, CROSBY AND KIRBY

LOST
COAL DISTRICT
OF
GEBO, CROSBY AND KIRBY

LEA CAVALLI SCHOENEWALD

Published by The History Press
Charleston, SC
www.historypress.com

Cover: These Owl Creek Coal Company (OCCC) miners worked in the 1930s in Mine #1. *Left to right*: Steve Patik, unidentified, Frank Pavlus, John Kowlok. *Kowlok family.*

First published 2024

Manufactured in the United States

ISBN 9781467156462

Library of Congress Control Number: 2023948344

CONTENTS

ACKNOWLEDGEMENTS

I wish to acknowledge the assistance of many individuals and organizations. First, Iris K. Guynn wrote a wonderfully crafted paper for her master's degree at Black Hills State College that discussed the interaction of the nationalities of Gebo, Wyoming, during the heyday of the coal mining industry. Her daughter, Amy Guynn Ready, gave me complete access to her mother's paper, as well as to her research. Both were invaluable.

Dorothy G. Milek's *Hot Springs: A Wyoming County History* contained invaluable information regarding the early days of Hot Springs County, as well as the various industries and people that helped make our county what it is today. Once again, her family allowed me to use any and all of Dorothy's extensive research.

Annie Kowlok Jones donated her personal Crosby archive to Hot Springs County Museum and Cultural Center (HSCMCC). She was a marvelous observer, writer, storyteller and poet. I needed her expertise time and time again. Her daughter, Marilyn Jones Revelle, happily added stories to the Crosby portion of this effort.

Dr. Donald "Duke" Bolich provided me total access to the Gebo portion of his autobiography. Not only did he explain the daily "Gebo life" beautifully, but his precious stories give such reading enjoyment as well.

The Wyoming State Historical Society awarded me two research grants through the Lola Homsher Foundation. These funds allowed me to travel, interview numerous people and gain a much-needed background for the story told here.

William J. "Bill" Deromedi was an invaluable resource. Not only did he provide numerous important stories about his growing up in Gebo, but he also provided a tutorial on Coal Mining 101. He and his wife, Ann, opened their home on numerous occasions and showed us many Gebo mementos. He also provided me with previously unknown information about my Gebo family.

Historian Ray Shaffer will forget more about Hot Springs County than I will ever know. He graciously agreed to read my manuscript and verify the historical facts I included. He also provided historical information that I was "missing" and hoped to include.

Jerry and Shelley Deromedi were our "first ride" to the district. They took us out and showed us around Gebo and pointed out various landmarks and abandoned mines. Shelley also shared her family's story from their time, and life, in Kirby.

Rudy and Linda Kowlok were also tour guides in Kirby and Gebo. Because Rudy lived in both, and actually worked in his father's mine, he offered a unique perspective. The photos he shared were also much needed.

So many people were gracious enough to share personal and family stories from Gebo, Crosby and Kirby, including Kelly Punteney, Janet Zupan Philp, Nena Roncco James, Barbara Heron Workman, Dolly Bury Daniels, Iris Carter Cavalli, John Ralph and his daughter Latrell Stevens, the Johnstone and Woolman family, the Pietila family and Mary Gordon.

Photographer extraordinaire Levi Shinkle contributed beautiful images. He allowed us to drive him around the remnants of the Kirby coal district on more than one occasion, and his remarkable eye captured our part of Wyoming with its unique beauty and history. He also visited private homes to photograph mementos from the district and proved a worthy hiking companion when visiting long-abandoned mines.

Connie Gifford-Kennerknecht, Dean King, Merrill Ready, Jackie Dorothy and the staff and board of directors of HSCMCC assisted me with access to the museum's research archive when needed, made a room available to complete research, gave me access to the majority of images included in this effort and scanned and resized those images.

The American Heritage Center at the University of Wyoming houses the Meliva Maravic papers. The collection provided numerous images of, and stories about, Gebo that would not otherwise be available for this effort. The staff was always willing to assist with research and answering numerous questions.

The staff of the Carbon County Historical Society and Museum in Red Lodge, Montana, and the staff of the Clarks Fork Valley Museum in Fromberg, Montana, provided fantastic information regarding Samuel Gebo's life before he came to what would become the Kirby Coal District. Becky Van Horn, preservation officer at Carbon County, was most helpful with providing Samuel Gebo's handwritten autobiography pages. She made every effort to locate needed photographs as well. At Clarks Fork Valley Museum, Marge Taylor (president) and Melody Kilwine (volunteer) were most helpful.

The Wyoming State Archives, the Wyoming State Newspaper Project and the Hot Springs County Library provided online and microfiche newspaper articles covering early Hot Springs County history. The Washakie County Library provided a microfiche reader, and the Washakie County Museum provided access to early oral histories of time spent in Kirby.

Heather Kroupa, Wyoming state inspector of mines, offered information regarding mining procedures and history. She was always willing to answer questions and offer encouragement.

The mayor of Kirby, Jessica Hurley, gave me my introduction to the town's historical archive. Town clerk/treasurer Sandy Scott and secretary Deanna Aylor were most helpful with answering questions about the photographs and town minutes. They were so vital in telling the Kirby story.

The Hot Springs County Clerk and Big Horn County Clerk offered access to early descriptions of various land transactions. This information made it easier to follow the progression of growth in the district. The staff in the Kirby Town Hall provided access to the minutes of the early Kirby town government meetings, as well as pertinent information about the town throughout the mining days.

My husband, best friend, chauffeur and cheerleader (and writing critic when needed), Thomas Joseph Schoenewald III, was an invaluable addition to this effort. His love of history and genuine interest in this project were, and are, so appreciated.

Unless otherwise described, all the images included here are found in the Hot Springs County Museum and Cultural Center photo collection.

INTRODUCTION

Coal mining is most definitely a part of my fabric. I am a coal miner's granddaughter. And niece. Coal mining even brought my family to America. Growing up, I was the oldest grandchild in Thermopolis, Wyoming. I got to spend the most time in the coal mining camp of Gebo (twelve miles north of Thermopolis) with my paternal grandparents, Bartolo and Maria Cavalli. Of course, my parents, siblings and I visited Nonna (grandmother in Italian) and Pa together. But when I was old enough, my mother put me on the train in Thermopolis, and I rode out to Kirby (three miles from Gebo, where the train depot was located) by myself.

Pa came to the depot to get me, and I went to the magic place that was my grandparents' home. My darling Nonna was an amazing cook, and she prepared everything on a monstrous black cast-iron stove. She deftly lifted the flat stove covers and threw in pieces of wood or coal. A pot of her marinara sauce simmered for hours over one of those covers. I was told by more than one Gebo resident that "all the kids wanted to play at the Cavalli house," in the hopes that my grandmother would invite them to stay for supper. To this day, every time I smell coal burning or prepare her "famous" sauce, I am back in her kitchen.

THE STORY OF COAL

Coal is not a mineral, because it is predominantly composed of organic molecules. These molecules contain carbon, nitrogen, hydrogen and sulfur. They combine to form compounds called "macerals." There are different classifications of macerals, depending on the type of organic material they came from, including trees, plants, fungi, seeds, bark, spores, roots and stems. Macerals actually determine the internal chemistry of the coal.

Coals are differentiated by "ranks," depending on the pressure, temperature and time they were formed under: lignite (soft, often brown, least amount of energy stored), sub-bituminous (black, not very shiny, harder than lignite), bituminous (black, shiny, significant amount of heat value) and anthracite (hardest, black, shiny, highest density). The coal at the center of this story is sub-bituminous, formed in the Paleocene epoch, Paleogene period, Cenozoic era.

There are also inorganic components contained in coal, either layers within the coal or single grains or crystals spread throughout the organic material. Within the layers can appear fine-grained sediment deposited during a flood or ash layers from volcanic ash "falls." Minerals can be found in coal, either primary (deposited at the same time as the organic matter by wind or water) or secondary (formed after the organic material was deposited).[1]

MINING

Throughout time, coal has provided humans with heat, light and a means to prepare food. The second Frémont Expedition of 1843 was responsible for the first discovery of coal in what would become the state of Wyoming. From that time on, "Wyoming" and "coal" have been inextricably joined.

According to historian Dorothy Buchanan Milek, the first mining claim in what would become northern Hot Springs County, Wyoming, was registered by Henry Cottle in the late 1880s. Shortly, there were claim stakes literally everywhere. The names on the claims included cemetery residents and other "ineligibles." Cottle worked his mine for a few years and then left for Ten Sleep, Wyoming, approximately forty-eight miles to the north, to look for more coal.[2]

Henry Monro and Frank Porter sank a shaft into Cedar Mountain, east of Cottle's claim, in 1898. It was christened the "Cowboy Mine." The coal was tested and determined to be "better steam coal than the Rock Springs [Wyoming] coal." It burned hot and left little ash. The vein, or seam, of solid coal was determined to be at least seven feet thick and at a favorable slope of twenty-three degrees. The Anderson family bought the mine and opened a coal yard in Old Town Thermopolis, which was already established because of the mineral hot springs.[3]

Around 1903, several mines began operating in the area: one was owned and operated by Prichard and his partner Fred Uriens, and a second was owned by "Dad" Eads and became the Highline Coal Company.[4]

In 1904, Cottle returned to the area and joined with T.P. McDonald to open the Cottle & McDonald Mine. Shortly after, "Dad" Jones started the highly successful Jones Mine. H.A. Stine incorporated the Jones Mine into the Kirby Collieries Company in 1907. A 125-horsepower hoist, boilers and an air compressor were installed. The first coal mined there was shipped to Cowley, Wyoming. Business really started to boom.[5]

Interest in mining coal in the district intensified with talk of the railroad eventually coming into the area from Montana. The Burlington Railroad made a monumental decision to build its southbound line from Billings, Montana, to Frannie, Wyoming, to Worland, Wyoming, in 1906. That decision was made, in no small part, because of the quality and quantity of coal near Kirby. The coal burned hot and left little ash (less than 3 percent). Another contributing factor to the Burlington decision was the "promise of tonnage" made by Rufus J. Ireland Sr. of Amityville, New York. He would figure prominently in the Gebo story.

The coal's quality would create a "problem" for the mines. Its quality was so high that the railroads charged "discriminatory" freight rates to haul it. Users were willing to buy lower-quality coal because the freight rates were lower and they could make a better profit. The positive reputation of the coal did eventually bring additional markets.[6]

That fall, Jesse W. Crosby Jr., a Mormon pioneer, obtained the contract for constructing the twenty miles of railroad grade from Worland to Kirby. Kirby became the end-of-tracks in 1907. Crosby and his business partner were then contracted to build a short spur from Kirby, southwest up Coal Draw, to the settlement that would eventually be called Crosby. He filed a mining claim on April 25, 1910, that consisted of 145 acres. Big Horn Collieries of Basin, Wyoming, purchased Kirby Collieries Company in 1910 for $250,000.[7]

More entrepreneurs wanted a piece of this action. Such a man came to the area in the person of Samuel W. Gebo (pronounced *JEE-boh*). As a teenager, he worked in the midwestern and western United States, and he acquired his own coal mine before the age of twenty. He was well known enough in Montana to become involved in the celebrated feud between his boss, Senator Clark, and Marcus Daly. He had heard about the quality of coal in northern Wyoming from his time working with the coal industry in Red Lodge, Fromberg and Gebo, Montana.

Eventually, within the district grew a true melting pot of nationalities: Italians, Greeks, Scotsmen, Irishmen, Finns, Serbians, Englishmen, Japanese, Welshmen, Turks, American common stock and others. Many of these groups experienced ridicule in their home countries and certainly when they came to America.

In Gebo, Crosby and Kirby, there were difficulties between the different cultures, certainly. Some of the countries had been at war with each other. But there was tolerance and true acceptance of each other as well. When the men entered the mines, it was understood that each miner had to depend on every other man (and they on him) as if his life depended on it. Because it did.[8]

William J. "Bill" Deromedi, who was born and raised in Gebo and whose father worked in the mines for many years, offered the following information about working in a coal mine.

Going to Work

Mining coal in the early to mid-twentieth century was a completely different prospect from what's done today. When mining in the district began, men worked underground with horses and mules. The animals were employed to move the coal cars around the "rooms" and then to pull the full cars back to the main slope and up to the surface.

Many people were concerned about the treatment of the animals in such harsh surroundings, so a barn was built to house them. When technology improved, steam-generated electricity replaced the need for horse/mule power. Electricity also allowed for the use of drills, which greatly increased the amount of coal that could be extracted each day.

All work of loosening coal from the vein and loading it into the cars was done by explosive charges, picks and shovels. As technology improved, the cars were pulled by a butt hoist or winch that was attached to the mine car into each room. Each room had its own butt hoist, or winch.

How to "Shake" a Coal Vein (after electricity came to the mines)

Undercut the area to be blasted, using a Cutting Machine, to a depth of seven feet
Drill holes strategically in the area, using a Hand Auger
Insert blasting cap in one end of dynamite charges
Insert fuse in blasting cap
Crimp blasting cap down on fuse
Insert dynamite in holes
Make "dummies," by rolling sand up in newspapers, crimping ends to hold in sand
Stuff a dummie in front of each dynamite charge, leaving each fuse loose
Insert all fuses into a plunger
Set off charges
Shovel loosened coal from the wall and into car

—*Bill Deromedi*[9]

The method of mining employed in Gebo and Crosby was called "room and pillar mining." It entailed "partially mining a vein (or seam), leaving large pillars of coal intact to support the overlying layers of rock." On either side of the main shaft, the mining engineer had previously determined how much space was needed between each pillar between the rooms. "This method created a network of alternating spaces and large pillars of coal."[10]

Eventually, the mines ran three eight-hour shifts, Monday through Saturday. In later years, the miners were given the entire weekend off. Men arrived at the mine and went to the gigantic (150-foot-long) shower house. There, each man had his own locker, where his dirty work clothes were hanging above the locker on a pulley and short chain (where they were hung after each shift to dry out the sweat). The clean clothes worn to work were placed in the locker, the dirty work clothes were taken down from the chain hoist and the men were ready to go to work.

Work clothes consisted of bib overalls or jeans, shirts, boots (later they were ankle-high rubber boots with steel toes) and wide leather belts. They had to wear lamps on their heads to give them more light in the dark mine. The first carbide lamps were attached to a canvas cap and used carbide gas and water to form an open flame. Later, the caps were made of metal. Because open flames were deemed unsafe, an electric lamp attached to a battery pack (which was much safer and put out more light) became necessary. The

Fig. 22—A Section of the Mine Working Plan.

A hand-drawn map of Mine #1 shows the intricate system of pillars and rooms.

battery was held by a carrying case with straps that a belt could be looped through. The battery pack had to be recharged after each shift.

In the earlier days, men and animals walked down into the mine together. Later, all the men gathered together at the mine entrance and got on the "mantrip" (a shuttle "train" on tracks, which also brought them back out). The coal cars that went down into the mine also rode on the tracks.

The shot fire boss (a mine official responsible for safety precautions) went down first, with a special lamp. If the air contained too much dangerous gas, the lamp went out and signaled that there wasn't enough good oxygen.[11] Because of the confined space in underground coal mines, adequate fresh air was necessary to prevent deadly gases (or damps) from building up. The most common of these were black damp (carbon dioxide mixed with other unbreathable gases, reducing the available oxygen and causing asphyxia), white damp (a mixture of carbon monoxide and other unbreathable gases, a "silent killer" and highly flammable and explosive), firedamp (a mixture of methane gas and other explosive gases) and stinkdamp (hydrogen sulfide, a flammable, very poisonous, rotten egg–smelling gas).[12]

Another danger faced by the miners was water. As they dug deeper into the earth, they encountered the water table. As the water flowed in more rapidly and became deeper, it also became dangerous and impeded the whole operation. It was constantly pumped out of the mines, but it was more and more difficult to keep up the desired production pace.[13]

Two miners shovel coal from a coal cutting machine in Mine #1. Their carbide-lit mine caps offer sufficient light to work by.

This metal mine cap contains a carbide lamp and belongs to Bill Deromedi. *Photograph by Levi Shinkle.*

Hundreds of men went through the entrance to OCCR Mine #1 every day.

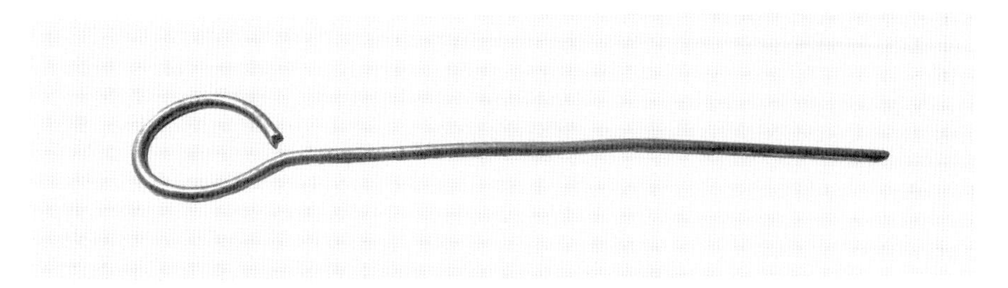

Above: Bill Deromedi fashioned this rope rider's pick to demonstrate its look and function. *Photograph by Levi Shinkle.*

Left: A miner's lunch pail contained a compartment for the miner's sandwich and so on. There was also a compartment for drinking water. This lunch pail is Bill Deromedi's. *Photograph by Levi Shinkle.*

The mine foreman assigned each two-person team their room for the day, which stayed the same until the room was mined out. The rope that was attached to each car could only be a certain length. When each mantrip reached a team's designated work area, the team was dropped off. Their car to be filled was already there.

The rope rider (the most dangerous job) was responsible for attaching the cars to the rope to be hoisted out of the mine. He also used a pick to send signals on the bell line to signal the hoistman. The bell line consisted of two parallel bare wires that were low voltage and would not cause an electrical shock. The rope rider manipulated the pick between the two lines to send a code to the hoist house. The codes included: one long and one short bell, pull up slow; one long and two short bells, let it down slow.

Each team was responsible for filling its car as quickly as possible. Contract workers were paid more as they loaded more coal. Company men were paid only a daily wage. They also laid track, set timbers (timbered up) under the roof to keep it from caving in and installed the bell lines.

The miners worked from start-of-shift until time for lunch. Contract workers took as little time as possible to eat their sandwich and drink some water out of their lunch pail. Company men didn't take much more time.

When a car was full and needed to be hoisted up, the team pushed the car to the main slope. This let the rope rider know the car was ready to be hoisted out. He attached the rope to all of the cars ready at any one time. The cars themselves were attached to each other by a clevis pin. The crew was given credit for its full car by means of a flat metal disc called a "check" that was attached to the front of the car. Each crew's check had a specific number that identified that crew. The hoistman then started the electric hoist, and the cars were pulled up.

COAL TO THE TIPPLE

Once the full cars reached the surface, they were pulled to the hoist house. The cars were weighed, and the check discs were removed and given to the business office. This gave each team credit for the cars they filled. There had to be a switch located on the tracks on the top to send the full cars to the tipple and the empty cars from the tipple back down to the workers.

Loaded cars wait on the tipple. *Mileva Maravic, Photographs: Gebo Mines and Employees, "Old Mine Tipple #1," 1919–1927, ah06309_0033, American Heritage Center, University of Wyoming.*

The tipple and the adjoining structures were necessary in the mining operation.

The full cars were sent on a downward-sloping track to the tipple to be unloaded on the shaker. The tipple was a huge rectangular building on timber stilts above five sets of railroad tracks. Each car that entered the tipple was sent to and emptied onto a long, flat piece of metal called a shaker. The shaker had five different sizes of holes and was moved slowly back and forth by a lever. As the different-sized pieces of coal moved over the shaker, they dropped through the holes into a corresponding bin below. From the bin, the coal was loaded into a boxcar, with each boxcar containing a different size of coal. Bill remembered three common sizes of coal, largest to smallest, as lump, egg and nut.

The pieces that were too small to be loaded into a boxcar were called "slack." The slack was dumped out of mine trucks into mostly four- to five-foot piles, one right after the next. The piles just kept appearing, going east and west from the tipple. Eventually, a few of them spontaneously combusted. One huge pile burned continuously for years.[14]

GOING HOME

When the mine whistle blew to signal the end of the workday, the miners rode the mantrip back out of the mine to the surface. At the end of the morning shift, the whistle blew either one or two times to signal everyone in the camp if there was work the next day. One whistle meant no work; two

The mantrip carried the miners into and out of the mine.

whistles meant work. A sign was also posted in the mine office window. The men walked back to the shower house and took off their dirty work clothes. These clothes were rehoisted above each miner's locker, and the men took showers. Clean clothes from the lockers were put on, and the men were done for the day.

CHAPTER 1
CROSBY

The Mine

Annie Kowlok Jones was born and raised in Crosby. She was the camp's historian and a gifted writer. As one of the many proud "Crosbyites," she provided the following history of her camp.

The founding of the coal camp of Crosby occurred in 1908, when the Big Horn Collieries Company of Denver, Colorado, bought a 160-acre lease from "Dad" Jones. He had seen great success during the years he operated his Jones mine. Besides the coal's excellent properties of burning hot and leaving little ash, it came from the same vein as Gebo's Mine #1; and it was determined to be eight to fourteen feet thick. The vein was situated at a pitch of eighteen degrees, in a northeasterly direction. The first superintendent of the mine was Dan Harrington. He was followed by Frank Anderson Jr., who remained in that position until 1933.[15]

At the beginning, mules and horses were used to power the machinery and to haul the coal cars in and out of the mines. The animals were kept in the mine full time. It was common for them to go blind, as they were never exposed to surface light. Many Crosby residents became concerned about the animals' welfare, and a large barn was eventually built near the mine.[16]

Big Horn Collieries strove to be known as a genuinely caring entity in other ways as well. The company collaborated with ranchers in the Big Horn and Owl Creek Valleys to provide electricity to residences in the

At the Kirby Collieries Company (KCC) in Crosby, the miners are shown at their various "positions" above ground.

immediate vicinity. It had the capacity to produce its own power and had offered to provide electricity to the agriculture community. In 1921, Crosby itself got electricity. In addition, the company brought in hogs to get rid of the rattlesnakes that were numerous in the area.[17]

Once fully operational, the mine attained a capacity to produce one thousand tons per day. It was equipped with modern (to that time) equipment, and it was known for its favorable working conditions:

> *It was electrically equipped throughout, having electric hoists, electric pumps, electric haulage locomotives and electric coal cutting machines. Ventilation, which was so critical in any mine, was provided by a "steam-fired fan," and the main coal haulage hoist located on the surface was steam driven. The power plant had 540 boiler horse power, two air compressing units, and two electrical units.*[18]

The company paid the miners twice a month. The exact figures of monthly payrolls were never made public. As in Gebo, a miner made a living wage, but there was little left for luxuries. Few people could afford cars, so walking

Crosby miners stand at the mine entrance with a horse and a child, which is highly unusual. Young children and women were not allowed to enter the mine.

or carpooling were the choices to get around. In 1924, the telephone lines between Crosby, Gebo and Thermopolis were cut. It was rumored that a payroll heist was being planned, and there was much fear in both of the camps. It never materialized.[19]

John Ralph, who lived in Crosby from 1927 to 1934, related a story he often heard: "One Monte [Montenegrin] got paid [converted his pay from currency] in gold and wanted to send it back to his family in his home country. In order to keep it safe, he buried it in the tipple area. For years, people swore it was still there. And wanted to dig it up."[20]

The day after payday, not all miners made it to work. "The sidewalks and streets were jammed on paydays and sometimes the poker games and drinking bouts would last for a couple of days. This is not to imply that all miners participated in such a manner, but as one fellow put it, 'they (the miners) worked hard and they played hard.'"[21]

The railroad, which was "staked in 1909 and built in 1911," allowed Crosby to grow, as the spur came up from Kirby to the Big Horn Collieries mine. The line of full coal cars took many minutes to make the trip back down to Kirby.

A steam locomotive waits to pull full boxcars from the Crosby mine to Kirby for shipment. The photo is labeled "Spotting Cars."

At least four men died in the Crosby mine. Many were injured, some permanently disabled. One young man died climbing over a moving train. Another lost part of a hand and a foot when he fell under moving wheels.

Like Gebo, Crosbyites were in tune to the signals from their mine's whistle. One spring day, Joe Zulevich was playing hooky from school when the whistle started its "emergency" blast:

> *Knowing that my Dad was underground coupled with my constant worry when Dad was underground I was terrified, so post haste I hurried to the blacksmith shop which was at the collar of the slope so I could see who they brought up. Dashing into the shop I almost ran over him. They had already brought him up. He was a huge Greek that I knew, laid out on a stretcher waiting for the hearse to come out from Thermopolis to get him. He looked like he had been run over with a steam roller. I hurriedly resumed my hookey playeing* [sic]. *But that night I couldn't sleep for thinking of that poor fellow.*[22]

A coal strike was called in 1921, and all the Crosby miners honored the strike. It lasted for nearly two months. It reached the point where the state militia was awaiting orders to move in and keep the peace. There were rumors circulating that scab (outside workers who crossed the picket line) labor was going to be brought in, but there is no evidence that ever happened.[23]

"MR. AND MRS. CROSBY"

One of the camp's residents described Crosby as

> *a beautiful little city nestled among the hills. It truly was just this. Not as much as the scenic beauty—there were some trees, a few patches of green lawns, some flowers—but the people themselves. Their kindness, consideration and compassion for each and everyone. All these things together left an unforgettable impression in the lives of all who lived there and the lasting, loving memories.*[24]

Joe Zulevich remembered:

> *Crosby was a very heterogeneous gathering of nationalitys* [sic], *who generally lived much according to their foreign customs. But they all had one thing in common, and that was the ability to feel the pains of their fellow man and would do all in their power to alleviate any such pain. With very rare exceptions Mr. and Mrs. Crosby were very fine people. I prefer to think of them as men and women of stainless steel with hearts of gold.*[25]

Because Crosby was never incorporated, an official census was never available. However, in 1920, the unofficial population was 494; in 1925, it was 600, with 250 men employed in and around the mine; and in 1930, it was 303.[26]

People lived in dugouts, tents and wood houses, and there were "suburbs" in the camp that reflected the different nationalities and social statuses of the residents: East Crosby, Tent Town (also known as "Rag Town" and "White City" because most people lived full time in tents stretched over wooden frame works in a basin in the northwest corner of camp), Greek Town/Greek Avenue (cabins built running east and west, both sides of the street, on the northern perimeter of camp) and Slack Town (people lived in dugouts on both the southwest and southeast edges of camp). Some of the bachelor miners also lived in dugouts dug into the north side of Coal Draw, below the family-oriented homes. It was rumored that they didn't spend much money. They also didn't trust banks, so they "buried it into the banks of the Draw." A few of the poorer families lived in boxcar-type dwellings. They had two "rooms" and a lean-to add-on. All of the houses were rented from the company for seven to fifteen dollars per month. A minimum amount was charged for utilities. The coal for home use was provided by the company. It was dumped into piles, one for so many houses, and then everyone had to scramble to quickly fill their coal house.[27]

In the warm months (after the mining season), when there was no work in the mines because people didn't need the coal for warmth, many families loaded up their camping gear and spent time in the Big Horn Mountains. As there was no money coming in, there had to be much creativity applied to their everyday living. Crosbyites ate fish three times a day, as well as an occasional poached game animal. Many of the men were employed by the company that built the highway through the Big Horns. If the windows in a Crosby house were boarded up because the family had left for a month or more, no utilities were charged for that amount of time.

Rose Barham Kitzerow grew up in Crosby and had many fond memories. She recalled that water reached most Crosby homes and businesses in "water mains." The water was pumped by electricity from the Big Horn River in

Crosby homes are pictured in 1919. "Bachelor dugouts" are also shown below the others.

Kirby. Those houses that didn't have water pipes got theirs from outdoor faucets that froze in the winter. Earlier, the water had to be hauled up from the river, and each family and business had to fill buckets. Because the water was "raw" and piped up to a storage tank near the tipple, all of the children had to get shots against typhoid fever every spring.[28]

Because many families could not afford to purchase meat on a regular basis, hunting was a common practice. Rabbits were plentiful in the Crosby area, so many meals included them. Even the children (both boys and girls) were adept with a rifle and helped their families by providing a meal.[29]

Rose laughed about the girls' wardrobe during her Crosby days: "When I think back on how we used to dress, it is rather funny, by today's standards. My sister, Grace, and I had to wear long johns from early fall until May first, along with socks. As I remember, the socks were black and we must not forget satin bloomers, the four-buckle overshoes, leggings, and heavy coats."[30]

Buildings and Businesses

The first Crosby post office was opened in 1908, with Harry A. Stine serving as postmaster. It was located in a dugout. The Big Horn Trading Company (the company store) began operating a general store and meat market. The post office moved to a corner of the general store when it opened.

The other businesses that were located in Crosby included a taxi service, first owned and operated by a Mr. Thomas; a shoe repair shop; a watch maker and "repairer"; a barbershop; several bars; a rug maker named A.E. Peterson, who owned a huge rug-making loom and "turned out beautiful rugs"; a "very talented" artist named Otto Hager; and a bachelor glassblower called "Little Frenchie," who "made many pretty articles."[31]

Residents could also shop at two grocery stores in Kirby, one run by the Nelson family and the other by Ferko and Tatarka, and at the Lucerne (also called "Spot") Trading Company operated by Bob Bender.[32] Families had accounts with the stores. When a family's bill from any of the stores was paid, a sack of candy bars was given to them—much to the delight of the children.[33]

Bob Bender had a beautiful singing voice. Because there was a lack of opportunities to listen to music, many Crosby women were "music starved." Their solution to the problem was to make Bob "sing for his orders. It was a common sight to see Bob standing on a door stoop with an order book in his hand, singing his heart out."[34]

Each of the Kirby stores made daily deliveries and took orders for the following day's needs. Milk deliveries were also made daily, first by Henry VanDusen (in 1918, with the housewives having to meet him with their pans and pails) and then by Can Reynolds, Arthur Gwynn and Al Graham. Fresh vegetables were delivered for years by the Palmer family of Worland. The iceman gave the kids a piece of ice to suck on, which took the place of an ice cream man.

Freight was hauled from Kirby by Russell James by horse and wagon. James was also the "barnmaster" for the mine mules.[35]

MEDICAL CARE

The camp boasted of a well-appointed hospital. It was run by a full-time nurse, who was ready to handle all types of injuries suffered at the mine. She also provided other care to the miners and their families. The physicians who staffed the hospital included Drs. Brantley, A.M. Giddings, Montgomery (whose office was in Thermopolis), Pierce and Nuttall (who were both doctors in Gebo). All of the medical supplies and medications were provided by the company.

Dr. Allen M. Giddings served as company doctor for fourteen years. According to Joe Zulevich:

> *Dr. Giddings was probably the most beloved and respected person Crosby ever had. Being a very highly proficient general practitioner, he performed many great medical feats for his time. Doc Giddings blended himself in with the community. He had an elegant sense of humor. He once had a horse tooth which he would palm in his hand, when a kid came up for a tooth extraction. When he extracted the tooth he would show the horse tooth. Doc was a good baseball player. He played third base and wielded a productive bat.*
>
> *In making home brew it was necessary to let the brew work a certain length of time before bottling then set in a bottle for some time before drinking. But often miners would get impatient and drink from the crock with cups. At times Doc would participate with the miners with this imbibement. Doc and his wife had three children. The two oldest being exceptionally brilliant scholastically.*[36]

Mrs. A.M. Giddings wrote a newspaper article in March 1925 stating there was "a small but well equipped hospital [in Crosby] with a trained

Dr. Allen M. Giddings was the beloved physician in Crosby for many years.

nurse in charge. The hospital is prepared to take care of injuries incurred in the mine, or other serious illnesses of the men and their families. Dr. A.M. Giddings has served as camp physician and surgeon for several years."[37]

Dr. Giddings was also remembered for how he cared for his patients, both physically and emotionally. Annie Kowlok Jones shared that her brother, Matt Kowlok, was burned in an explosion at a gas station in Ten Sleep, Wyoming. Dr. Giddings left Crosby, and his patients (with the Gebo doctor), to stay with Matt. He stayed with his patient for a week and offered comfort to both Matt and his family.[38]

The Giddings family was very involved in the Crosby community. Dr. Giddings was a member of the school board, and Mrs. Giddings served as superintendent of the Sunday school.[39]

Schools

Long sat the schoolhouse by the track,
Filling with memories of the present and past,
Forming of character, building of man,
Molding of friendships to last a life span.
—Annie Kowlok Jones[40]

Having a local school was very important. The first school district that included Crosby also included Lucerne, Gebo and Kirby. It was divided and became District #46 in Big Horn County, and it included just Crosby, Gebo and Kirby. Hot Springs County was formed in 1913, and each of the schools formed its own district. Crosby was named District #3. In 1925, the enrollment was 125, and 3 teachers were hired. The high school students attended school in either Gebo or Thermopolis, and the company paid the tuition costs. The company also covered all books, teaching supplies and teachers' salaries. The students brought their own tablets, notebooks, pens, pencils, ink and crayons.[41]

Crosby School was first located in a residence. Later, a small one-room home in the central part of the camp was used for the primary (lower)

Miss Lucy Coleman and her students stand outside Crosby's "little school," circa 1914–15.

grades. It was called the "little schoolhouse." Kindergarten was included for several years. The upper grades (through eighth grade) were housed in a converted two-room warehouse, the "big schoolhouse." Both schools were heated with coal stoves, and the big schoolhouse had a lavatory and running water in one of the rooms.

Annie Kowlok Jones reminisced:

> *All the basic subjects were taught—Reading, Writing, Arithmetic, History, Geography, Grammar, Spelling, Health, Art, Agriculture—there was always homework to do, many poems to learn, books to be read for reports, and paintings to study. Singing was from the Golden Book of Favorite Songs. A fifteen minute recess, supervised by the teachers, was twice a day. There was always a race for the two swings, the teeter-totter and rings and the slide. Group Games, marble playing and ball playing were part of the recess and preschool activity. No playground equipment was at the primary school.*[42]

The big schoolhouse had been built only ten feet from the railroad track for "shipping accessibility." The switch engine came up from Kirby in the late afternoon to "switch the mines" and pick up the full boxcars. When the engine was going by, there could be no reciting; the noise was terrific. The full boxcars were pulled back to Kirby, sometimes taking forty-five minutes to pass by. The students couldn't doze off in school, as the noise was still evident.

If the engine and cars passed by during recess, the kids threw objects at the brakemen as they walked the top of the cars. That angered the men over the years. The students also placed pennies and nails on the track so the engine wheels would flatten them. The engineer or fireman would sometimes open up a valve and "blow steam" so the students couldn't get close to the track. Occasionally, the more brazen boys jumped on the footboard of the engine and rode up to the store.

Rose Barham Kitzerow remembered having to leave early in the morning for school. The children carried their lunches in five-pound lard pails. In the winter, everything was nearly frozen by the time they reached the school building. That was not a problem, as they all set the pails around the big pot-

Class is in session at the Crosby "big school."

bellied stove. The lunches were all thawed out by lunchtime. Rose's first and second grade teacher was Mrs. Hill, who was very strict and didn't hesitate to "use the strap" when she felt it was necessary.[43]

It was a "big day"

> *when we passed the second grade and went to the big schoolhouse. Mrs. Garrison taught the third, fourth, and fifth grades. We knew when we were in her room there was no foolishness, that we were there to learn the three R's. We thought at the time that she was really mean, but some of the things I learned in her room I have retained throughout the years.*[44]

When Rose was in the sixth grade (in 1929), her entire class became very interested in track. Their team became very successful and won many pennants and silver cups. Rose participated in the relays, 50- and 100-yard dashes, hurdles, baseball throw, discus, shot put, high jump and broad jump. Her teacher, Miss Anderson, insisted that the track students train at school. They ran around a track that was equivalent to a mile. Then, they practiced their individual events. Shorty Raybould, the uncle of Rose's friend Myrtle Anderson, helped coach the girls' team. Two of Rose's friends, Myrtle Anderson and Elsie Toth, often stayed overnight at Rose's, especially if they were training for track. They got up and ran a mile or more before school after Rose's mother gave them a "good breakfast."

The girls' track team members from Crosby show their winning form. *Left to right*: Agnes Panetus, Rose Plachy, Janet McClusky, Rose Barham, Elsie Toth and Myrtle Peoples.

Annie Kowlok Jones and her classmates relax outside the "big school." Annie Kowlok is in the back row, third from the right. *Annie Kowlok Jones collection.*

The end-of-season track meet was held in Thermopolis. Some type of carpool was set up for the students to travel into town. Their teacher rented a room at a hotel for the students to go back for rest and to change clothes. Their parents usually gave them a dollar to spend, and the students felt rich. Very seldom did they have that much money to spend as they chose.[45]

The school facilities weren't sophisticated, but the education that the students received was excellent. The various schools competed against each other in various contests, both academic and athletic. Crosby always won its share of awards. A favorite joke came from a Gebo teacher, who said that there were more dogs in the Crosby school than students. During cold weather, the dogs, when allowed, slept around the coal-burning stove in the classroom.

The school was open until 1933, when the Crosby mine closed. The students attended Gebo school for one year, and then the district was annexed and became part of Lucerne District #11.[46]

FAITH

Faith was certainly important to life in Crosby, and a Community Sunday School was held each week. B.G. Rodda, who managed the Big Horn Trading Company, conducted services on Sunday. Mrs. Rodda directed a youth choir and taught Sunday school for a time. The Sunday school officers were Mrs. A.M. Giddings, superintendent; Mrs. P. Diemer, assistant superintendent; Mrs. E. Bamber, secretary treasurer; and Miss Frances Richardson, pianist.[47]

Mr. and Mrs. Rodda's son George, and his wife, Gertha, hosted many parties. Halloween parties, especially, were popular for the younger crowd. The Sunday school also sponsored many activities for the young people in camp.

Other groups held their services throughout the week, making denomination choice available. Father Endres from Thermopolis said Catholic Mass on the first Sunday of each month. Catholics from Gebo attended this service as well. When Father Endres retired, his replacement, Father Krass, offered Mass in Gebo.

Rose remembered "a lot of evangelists coming through town and they would hold revivals at the hall. Everyone, or most everyone, went and if there wasn't room inside they would stand on the outside. If my memory serves me right, the revivals were more the 'hell, fire, and brimstone' type which scared us kids."[48]

Camp "Services"

There was no county-provided law enforcement. However, Pete Sparks was appointed as the constable when needed, and John R. Jones served as justice of the peace. Bob Ralph served as a "volunteer safety officer in the Crosby mine."[49]

Pete Sparks was Crosby's first appointed constable. *Smith family.*

The camp was home to its own version of "outlaws," including One-Eyed Whitey, Deacon Jones, Taxi Nick, French Pete, Crazy Ben, Mooch Bretthauer and Fat and Bigum Jones. Only one murder and one suicide could be remembered as occurring.[50]

Even though One-Eyed Whitey was considered an undesirable by some, he also helped the school students with their math homework after school. He seemed to be waiting for them in the company store, where they would spread their books and tablets on a glass showcase. He was "brilliant" and showed the kids how he could "add five rows of figures at one time."[51]

An organized fire department also did not exist in the camp; volunteers helped man the fire wagon. Some houses and tents were destroyed by fire over the years. The tipple was destroyed in about 1919. The damage was so extensive that the men were laid off and had to rebuild the tipple before production could be resumed. The bunkhouse was destroyed in the mid-1920s.[52]

ORGANIZATIONS

The Crosby Sick and Accident Society, officially known as the Crosby Mutual Benefit Association, was formed to assist the miners. Such entertainment as boxing, singing, smoking, card playing, meals and others was available. The kids knew them as "smokers." The society also sponsored community entertainment, featuring local talent. In 1925, the officers were: president, J.J. Jones; vice president, Fred Paulin; financial secretary, Thomas Donaldson; recording secretary, Vivian Dorius.[53]

The labor unions in Crosby and Gebo sponsored celebrations on Labor Day and on April 1. They brought everyone in both camps together for a common cause.

The women in the camp had their own group as well, the Women's Benefit Association (WBA). It provided social opportunities and insurance coverage to its members, and its mission was to make camp life more enjoyable for all the residents.

Mrs. Giddings reported in the *Crosby Review* no. 14 that the WBA was

> *an active organization, holding regular meetings twice a month, with Mrs. John Kowlok as commander. These meetings are social as well as fraternal and many good times are enjoyed by the members. Refreshments are served and entertainment provided for the social hour which follows every review meeting. The Junior Maccabees hold meetings under the direction of Mrs. A. Peterson. Ercil Thompson has recently been elected queen of the rose court.*[54]

On Thursday nights, Local 2700 of the United Mine Workers of American held its weekly meetings. This was the official miners' union and sponsored the society. The societies in Crosby and Gebo were responsible for two large celebrations that occurred each year. On April 1, Crosby hosted the commemoration of the eight-hour workday labor law. On Labor Day, a similar celebration was held in Gebo.

Entertainment and Fun

The celebrations were all-day affairs. The day was spent enjoying games, races and various contests—climbing a greased pole or catching a greased pig, to name a few. The event of the day was the tug-of-war between Crosby and Gebo. Prizes were provided by the two miners' locals and the merchants in Thermopolis. Sometimes, a baseball game was played. A big dance was always held in the evening to close out the day. Refreshments of all kinds were served. A delightful time was had by all.

The union also provided the camp's Christmas celebration by "playing Santa Claus." Rose Barham Kitzerow recalled the big schoolhouse became one huge room with the doors between the classrooms opened. A large tree was placed at the end of the sixth-to-eighth-grade room. The evening began with a Christmas program provided by the students, with almost all of the children having a speaking part. Then, each child received an "outstanding" gift and a bag of fruit and candy.[55] In 1920, Annie Kowlok Jones and her three brothers each had an orange under their Christmas tree. Annie also received a papier mâché doll. That was the extent of their Christmas that year.[56]

Even though the location of Crosby didn't provide many places for activities, there were still chances to have fun. Baseball was very popular, and the camp had a team almost every year. The playing field was nothing to brag about, but the teams were well supported and won many games on a regular basis.

The ice skating "rink" was Coal Draw, the wide ditch that ran through the camp. Sometimes, the skating was easy and fun. But other times, the skaters had to be on the lookout for alkali spots, which could slow down the skating and even cause accidents. Rose loved to skate:

> *Most of us had the old clamp on skates which were hard on shoe soles, they were probably worse on our folks than on us because money was not so easy to come by then. Sleigh riding was always exciting. If a person had a flexible flyer sled he was really distinguished. Some of us even got brave enough to make skiis* [sic] *out of barrel staves. As I remember some of the boys would play a form of hockey and they used a part of the old Model T Ford for their sticks. Charley Malloy was one of the better skaters*[;] *he was one of the fortunate ones to have shoe skates. We all thought it was pretty great that he could skate backwards.*[57]

Often in the evenings, families gathered at each other's homes for "kitchen sweats," as they were called. The men moved the furniture outside to allow more room for dancing and visiting by the adults. Music was provided by musicians who showed up and played the fiddle, mouth harp, guitar, accordion (often played by Johnny Bobeck) and banjo.[58]

The children enjoyed games, including red light, green light; pump, pump pull away; giant steps; kick the can; run, sheep, run; ante over; and others. Mothers and other women who attended brought sandwiches, cakes and coffee, which were served at midnight. Rose's granddaughter Karla O. Christensen added additional information for the story in a paper she wrote for her history of Wyoming class: "Taffy pulls, popcorn popping, and fudge making were also big hits. At a certain point, a group of older boys began presenting feats of hypnotism at various gatherings."[59]

A yearly occurrence that everyone looked forward to was the appearance of the Raleigh man, a traveling salesman. He sold a variety of goods: red liniment, various spices, anti-pain oil and "carbo salve" (phenol or carbolic acid), camphor, oil of spruce and tea tree relative cajeput.[60]

Rose's father told a story about a traveling salesman and the illegal alcohol so many Crosby men made. Nicknames for the incredibly potent concoction included home-brew and Dego Red. Karla included it in her paper as well:

> *He peddled milk and meat to the foreigners because it seemed they always had brew. One day he went to this one woman's house, she was known as Annie.*
>
> *"Good day to you Annie," he said. "Brought you some meat. Hey, some brew would sure taste good."*
>
> *"I don't have any bottled, but you're welcome to that brewing," she replied, and inside they went.*
>
> *Home brew was kept in crocks behind the wood and coal ranges where it was warm to aid in the fermenting process.*
>
> *There was a dirty dish towel covering Annie's large crock, she lifted it, there was a duck swimming around in the beer!*
>
> *"Annie," he exclaimed, with a look of repulsion on his face, "I don't think I care for any brew today."*
>
> *"Oh, but Will, he's a good duckie, he only swims in beer!"*[61]

A large entertainment hall was built onto the pool hall, and roller skating was enjoyed there. The roller skates were rented to those who didn't have them. The skates fitted on the soles of the shoes and were tightened with a key.[62] Movies and plays were also shown and given there. During the school year, the kids could "walk two miles over a muddy and rutty road to Gebo to see a movie for twenty cents."[63] A tennis court, a miniature golf course and a croquet court were enjoyed for many years. *Uncle Tom's Cabin*, *The Johnstown Flood* and Tom Mix "shorts" were popular in their day.[64]

Football was a sport that was surely enjoyed by players and spectators. A boy from Wales, Tommy Jones, did not understand the U.S. style of the game, as he was used to the "football" (soccer) played in his country. One game saw Tommy catch the football and then just stand there. He yelled, "What do I do with the blinkin, bloomin thing?" He was promptly tackled![65]

In the early 1930s, the Great Depression was taking a toll on the services that were available in Crosby. The electricity in homes was shut off from 10:00 p.m. until the next morning. Everyone was forced to use gas and kerosene lanterns. One Crosbyite complained, "Never could finish listening to Amos and Andy on the radio."[66]

THE END

In January 1933, the mine was closed early by Superintendent Frank Anderson. It was planned that the mine would be opened again in the fall, with the possibility of hiring a smaller crew than had been working in the

An explosion on January 16, 1933, permanently closed the Crosby Mine and caused an overwhelming fire that burned for decades. *Annie Kowlok Jones collection.*

previous season. There had been a fire in the mine since about 1912, and it had broken out again. Because the company had already invested a great deal of money trying to keep the fire down, and the season was coming to an end, it was decided to close early.[67]

On January 16, 1933, an explosion occurred at the mine that caused it to be closed permanently. All of the equipment was sold, most of it to Fred Goodstein from Casper. The buildings were sold to partners Adams, Dunn and Tippetts, who in turn resold them.[68]

Some of the homes were purchased from the company and were moved to other locations in Thermopolis and Kirby, including farms and ranches. As the buildings were being dismantled, the Crosby boys were offered five cents per day to pull nails out of the leftover lumber. John Ralph "took advantage of that opportunity."[69]

The fire from the explosion was visible for years. John reported that its oxygen supply was finally cut off in the late 1930s. From 1938 to 1940, the company tried to reopen the Crosby mine. John's uncle Bob insisted that reintroducing oxygen into the mine would only reignite the fire. But the company proceeded with its plan. The fire did indeed reignite and burned for decades after.[70]

As an adult, Rose Barham Kitzerow reflected:

> *All that is left of Crosby are memories kept alive by those who lived there. In order for that era we loved to survive, we must commit those memories to paper to leave for our children and to those interested in history. Strong bonds of friendship were formed that have continued over the years. I will treasure my memories and friendships forever.*[71]

In 1970, Annie Kowlok Jones and her daughter Kathryn Jones went out to visit the remnants of Crosby. They walked around to all of the house foundations, which Annie could still identify. When they came to the Kowlok house location, something shiny caught Annie's eye. Sticking out of the ground was a sterling silver pencil that her parents had given her in first grade as a reward for good grades. It still wrote.[72]

CHAPTER 2

THE PLOT

With the certainty of the Burlington Railroad choosing to locate its route through Kirby, then on south to Denver, demand for the wonderful coal in the district continued to grow. Samuel W. "Sam" Gebo finally decided that it was the right time to "make a significant move" to corner his portion of the potential wealth contained in the district.[73]

Sam was born and raised in New York State in a French Canadian family. He always prided himself on working hard, getting what he wanted and living well. He knew and believed that making huge sums of money by mining coal was his destiny.[74] (See appendix.)

He also realized that connections in the right places made that effort infinitely easier. His family was still living in New York State, and he had cultivated friendships with many wealthy businessmen there. He owned or co-owned several coal mining interests in Montana, and he entered into coal "concessions" with groups in France, Bulgaria and European Turkey.[75] He worked tirelessly to further the career of Montana U.S. senator William A. Clark Sr., who was heavily involved in mining, banking and railroads.[76]

Sam believed in the necessity of proving the "quality of the product." He personally paid to have four hundred tons of coal from near Kirby freighted 135 miles (and 110 miles from a railroad) to be tested. He declared that it was "the finest coal in the west."[77]

However, in 1906, there was a limit of 160 acres of government land on which any one individual could file a coal claim. Sam knew that this small amount of land could never yield the kind of revenue he envisioned. He hatched a conspiracy that would drastically increase the claims he and

Sam Gebo "figured large" in the early days of the district. *U.S. Passport Photo, National Archives.*

several associates could control—a conspiracy that was highly illegal. Even the *New York Times* would cover the story years later.[78]

Sam had partners in both New York State and Wyoming. The New York group included "Wilberforce Sully, Chairman of the Board of the American Malt Corporation; George W. Dally, a stenographer in the office of Alfred Sully, his brother; Rufus T. [*sic*] Ireland, a large property owner of Amityville, L.I.; and Frederick T. Wells, a lawyer, of Amityville." John Nelson, John B. Wight and Thomas P. McDonald "handled" details in Wyoming along with Sam.[79]

The *Times* reported:

> *In 1906, Wilberforce Sully, along with Ireland, Dally, and Wells joined in a plan to take out valuable coal land and got various residents of Amityville and a number of men and women in Harlem to make out the necessary affidavits declaring that they desired to take up 160 acres each of coal lands which the Government proposed to distribute.*

Many of the men in Harlem were bartenders.[80] Regardless of their social status, all of the names were considered to be "entrymen" (people entering into the agreement for the property in question) in the book of mining claims applied for.[81] "The alleged conspirators, it is charged, then organized the Northwestern Coal Company and the Owl Creek Coal Company, and gave certain shares in these corporations in exchange for the rights to the coal lands, paying, however, $5 for each affidavit made in violation of the Federal statute."[82] This intrigue resulted in the obtaining of rights "vastly in excess of the amount which any four individuals or any combination of individuals could take out."[83]

Once these declarations of desire to file claims were in place, it was time to get the ball rolling in the West. Using the newly formed Northwestern Coal Company, the partners were able to obtain 1,760 acres more than they were allowed to acquire. And through the newly formed Owl Creek Coal Company (OCCC), 4,312 acres were illegally obtained. "Each individual was allowed to take out 160 acres at $15 an acre, and the Government was actually paid the $15 rate on all the land covered by the rights which Sully and his associates acquired."[84]

In order to pay for the "con claims," a bank draft for $120,000 was registered in Lander, Wyoming, around June 13, 1906. Unbeknownst to anyone but the co-conspirators, and with working capital now secured, the illegal mining activity could commence.[85]

Regulations, however, had been written to prohibit such a scheme. In addition to these regulations, other forces were at work to prevent the elaborate con from succeeding. The astronomical amount of the bank draft, coupled with the large number of entrymen, caught the attention of the federal government.[86]

Additionally, there were entities in the eastern United States who believed that "not only should the west be an unpopulated preserve, but it should be administered from afar." A movement to "preserve forest land in the west" was championed by two brothers, Gifford and Amos Pinchot. Their influential beliefs certainly caught the public's attention. Sportsmen, in particular, saw the preservation movement as a way to "make game and fish available to the person who could not afford to own a preserve."[87]

There was one other possible clue to the origin of the government's knowledge of the con. It was believed at the time that big business tipped off the government about the fraudulent claims. Specifically, the Union Pacific Railroad and Northern Pacific Railroad, which both would have benefited from the Gebo claims and suffered from competition from the Burlington Route and the coal it carried, may have had a hand in the notification. The Theodore Roosevelt administration began a thorough investigation.[88]

All of these forces joined to ensure that Sam and his co-conspirators would be punished. By 1908, Sam was in litigation regarding the questionable entries. The *New York Times* continued:

> *After the discovery had been made in Washington that the plan had been worked, the case was sent over here and put in charge of Assistant United States Attorney Smith, who obtained an indictment from the Federal Grand Jury on the plain charge of defrauding the Government. Certain legal contentions were advanced by Louis L. Delafield, counsel for the defendants and for the Owl Creek Coal Company, which make it advisable to have a conspiracy indictment found in Wyo.*[89]

Sam left the United States and took up residence in Guatemala. He returned to the United States and was involved in various mining and other enterprises. He retired to Seattle, Washington. On July 10, 1940, he took his own life through natural gas asphyxiation.[90]

Even though mining operations were well established in Gebo, the federal mining office eventually decreed that the "administrative proceedings were to revoke the mining patents and enjoin the operation of the mines. Northwestern Coal Company and the Owl Creek Coal Company mines should be closed."[91]

By 1912, Northwestern Coal Company had been dissolved, and there was no question that the federal government was going to succeed in closing Owl Creek Coal Company. However, the UMWA local attempted to intervene by writing a letter to one of Wyoming's U.S. senators:

> *UNITED MINE WORKERS OF AMERICA*
> *Gebo, Wyo.,*
> *February 6, 1912.*
>
> *Hon. FRANCIS E. WARREN,*
> *Washington, D.C.*
>
> *DEAR SIR:*
>
> *I have the honor of writing to you to-night, being authorized by local union No. 2671 of the United Mine Workers of America to do so. We are very much concerned, as a body of men, over our work. No doubt you will think it strange that we write to you over this matter, but I will try to explain, and I have no fears but what you can help us. We are mining coal for the Owl Creek Coal Co. at Gebo, Wyo. I understand they are having some trouble with the government in getting then-patents; in fact, from what we hear, they cannot get them on account of making their first filings illegally. I have no doubt but such is the case, but it is not for us to say. All we have to say is that we are glad to work for them, and would like to see them get their patents or a lease or something that would allow them to continue work. We understand Secretary Fisher is about to make a decision on their filings, and have heard it said that they will lose out. In this event, we will suffer the most, for if he decides against them, they will be shut down, and we will be thrown out of employment in the dead of winter. There are several hundred people depending entirely upon this mine for a living, and most of them are men of families. It will be a serious thing if the mine shuts down, for the work has been very poor for several months and we are barely making a living now. But we don't know where to go, because other places are worse, and besides we have not enough money to move away. Now, we*

do not want the Government to condone any kind of fraud for our sakes. In fact, we are glad to see them take the stand they are in regards to some who have been violating the Nation's laws. But we feel that some leniency could be shown. If they cannot get a title to their land, can they not get a lease? Or if neither of these, could we not prevail upon Secretary Fisher to postpone his decision for a few months until the weather is warmer and we are in a better condition to move? It would be useless for us to write him, for it would have no weight, but he would listen to you and no doubt consider our conditions. I do not desire to weary you, but trust you will intercede for us, and you will have the best wishes of the entire community. Hoping you will do your best and wishing you success, I am, Yours, respectfully,

A.H. PETERSON,

Secretary, Local Union No. 2671,
United Mine Workers of America,
Gebo, Wyo.

Senator Warren did intervene on Owl Creek Coal Company's behalf, and Secretary of the Interior Fisher heeded the senator's request. Congress passed a resolution authorizing the mining property to be leased to the coal company. It was a positive situation for all the parties: the government received a royalty on each ton of coal mined, the company stayed in business and the employees kept their jobs.[92]

It was discussed in various circles that if the defendants had merely set up a number of corporations that had filed claims of 320 acres each, those filings would have been legal.[93]

CHAPTER 3

GEBO

The Mines

Sam Gebo's statement regarding the quality of coal in the district was prophetic. For the first several years, the increase in the tonnage mined increased dramatically on an almost annual basis. The insatiable need for coal on the part of railroads, homes, businesses and the World War I effort kept the orders for Owl Creek Coal Company's product pouring in for many years. The following information was provided by Rufus J. Ireland Jr.:[94]

Year	Tonnage	Number of Men Employed
1906	3,725	
1907	15,231	(1st car loaded August 27)
1908	72,018	130
1909	132,577	180
1910	114,137	135
1911	120,124	100
1912	120,699	90
1913	163,355	154
1914	174,611	160
1915	210,111	213

YEAR	TONNAGE	NUMBER OF MEN EMPLOYED
1916	261,790	243
1917	287,115	265
1918	320,343	282
1919	266,499	277 (a 30-day strike this year)
1920	348,100	390
1929	500,000	640

However, while the tonnage and number of employed miners increased during the years from 1906 to 1912, the fraud case and all its implications were dragging on.

> *Criminal cases against Gebo's coal entrymen were finally set for trial in Cheyenne, Wyoming on May 19, 1913. Some 40 witnesses were subpoenaed to come from New York and other places. It seemed to be conceded that had there been a series of corporations each taking 320 acres, rather than as individuals, no objection could have been made. After years of discussion, Gebo and four others paid fines of $1,500 each in the U.S. Court at Cheyenne. The $7,500 and costs were paid in May, 1913.*
>
> *Thousands of dollars that would have been spent on the townsite and development and machinery, were thus diverted to travel and lawyer's fees, and fines. Ingenuity that would have been expended in designing and developing the ideal townsite and mine and power and industrial development, was wasted on worry.*
>
> *During the years proceedings were pending, Eastern investors were wary of going to Gebo, into the jurisdiction of the Wyoming courts, to confer with the operating employees. They met at Billings.*[95]

The eastern investors were found guilty of fraud and other charges, but Ireland, Gebo and other investors were allowed to lease the coal lands (and pay the government a royalty on every ton of coal mined) and remain as stockholders in OCCC. Rufus Ireland Sr. even became president and principal stockholder of the company and president of First National Bank of Thermopolis. He was active at the mine for years, and his son Rufus J. Ireland Jr. eventually spent time in Wyoming as well.[96]

Rufus J. Ireland Jr., in a letter to Thomas A.L. Nicholas in 1968, stated that Rufus J. Ireland Sr. "organized the Owl Creek Coal Company, and received

OCCC Mine #1 was so profitable that the company opened a second location, Mine #2, in 1924.

the first government coal lease in the United States. When I gave up the lease in the late thirties, we had a little celebration in Fred Johnson's office in Washington. Fred was then Commissioner of Land Management."[97]

According to Mary Hanssen, Gebo resident and teacher at Gebo School (and later teacher in Thermopolis):

> *At its peak, there were 700 men with a payroll of $300,000 every two weeks. I estimate and base my assumption on the information that Ireland pulled out in 1938 and Pat Burnell ran No 2 Mine (which had opened in 1924) employing about 100 men; his payroll was between $40,000 and $50,000 every two weeks.*[98]

Clarence A. Barnard served as superintendent of the mines from the beginning. He was also elected to serve as one of the first county commissioners of the newly formed Hot Springs County. In 1920, he and his wife were killed in an auto accident at Cover's Cut. Going back to Gebo from Thermopolis, their car struck the guard rail and fell onto the train tracks below. He was replaced by James Rae. During Rae's tenure, the

OCCC built a superintendent's home on the hill near the hospital. The last superintendent was Pat Burnell.[99]

Even though the coal from Gebo and Crosby came from the same vein, there was still friendly competition between the camps. Gebo miners insisted that Crosby coal "had more slate." The Crosby mine purchased large advertising stickers to be pasted on the bigger lumps of coal. Passengers on trains that passed Crosby coal trains were "treated" to messages about the superior quality of the Crosby coal. Gebo miners answered, "It is good they furnish paper to make their coal burn."[100]

The mine communicated various messages to the miners and the camp. At the end of each workday, everyone listened for the signal. One blast from the whistle meant there was no work for the following day; two blasts meant there was work. Another message came from the coal cars themselves, plus the whistle:

> *Perhaps the most frightening sound at the camp in Gebo was the sound of the cars slowly ascending out of the 6,000 foot deep mine. This meant trouble of some sort, either an accident or a death in the mine. The people attuned themselves to the sound of the cars coming out of the mine with a speed and force that was easily recognizable. But when the cars moved slowly and the long wail of the siren broke the calm, there had been a*

Left: John Bury Sr. is shown with his mule at OCCC Mine #1, circa 1930. *Bury family.*

Below: A cave-in (this one in Mine #1) would always cause alarm.

Left: Rufus J. Ireland was the main financier of OCCC. He made significant contributions to Gebo and Thermopolis for many years.

Right: From Barnard's death in 1920 until the early 1930s, James Rae served as Mine #1 superintendent.

> *disaster and the people were terrified. Although the regulations stated that only those employed in the mines could be at the entrances of the mines, all came running to find out if there had been a cave-in, and/or who had been injured or killed. Children and teachers left school, wives ran from their homes, and the shopkeepers deserted their businesses to discover if the miner was a relative or neighbor. Mining people were a conglomerate group as far as ethnic backgrounds were concerned, but disaster brought them together, and they cared for each others* [sic] *family and possessions until the tragedy had passed.*[101]

THE MELTING POT

The demand for coal dictated the number of days the miners worked in any year. During the warm months, there were far fewer workdays, as coal was not needed to heat homes and businesses. In some years, fewer than ninety days were available to make a living. Many miners made less than $500

Above: The coal camp of Gebo had very humble beginnings, circa 1905–10.

Opposite: A fishing camp similar to this in the Big Horn Mountains was home to many Geboites during the summer. *Left to right*: Nick Payovich, Charlie Flagg and (?) Pavkovich.

annually.[102] This amount meant very little, if any, money for extras: very few families could afford a car, so people walked most of the time or carpooled; clothing was handmade; and many families had boarders living with them or provided bachelor miners with meals and laundry services. Other miners worked on farms and ranches, in the oilfield and on construction sites. "They had few resources, but they were resourceful."[103]

There were two types of work: company work and contract work. Each individual miner decided which type of work he did. Company wages were about $3.50 per day. These men laid track, placed timbers to shore up the ceilings in the mine and did similar tasks. The contract workers were paid by the number of carloads of coal they loaded in a day. In order to get credit for each carload, a check (a flat metal disc with the crew's number on it) was attached to each coal car the team filled. That way, the team was given credit for that car and was paid accordingly.[104]

A.M.O.S

Many families packed up and went to the mountains after the coal "season." Getting out of the hot Gebo summer was desirable, and fishing in the lakes and streams provided free food. Illegal hunting of game provided additional meals, plus surplus wild meat and fish were preserved to be eaten during the colder months. Many of the Gebo miners made extra money by building the new highway over the Big Horn Mountains.

Occasionally, a contract would come to the mines that required a larger crew. The men came down from the mountains to earn the extra money. Families left their houses unlocked while they were away, and they made the trip home to check on their property every so often.[105]

In July 1971, Iris K. Guynn submitted a comprehensive paper to Black Hills State College in partial fulfillment of the requirements for the degree of master of education. Her topic was "The Social Interaction among the Residents of Gebo, Wyoming." The following descriptions of nationalities are taken from that paper.

Greeks

The Greeks who came to the mines were single men. They came to make a better living and then return to their native land. They were very proud and were not willing to admit they immigrated because they were poor, but they were willing to do just about any task to make money. They were mostly frugal and were able to save significant amounts of money. They came to America as chaste individuals, but that chastity did not always last.

For the majority, the Greek Orthodox Church teachings were followed. The Greek priest was close enough to the area to be available for births, weddings and burials. He was respected by his parishioners but not feared. In fact, many followers considered their priest a true friend.

The Greeks were extraordinarily proud of their heritage. In fact, it was common to understand that Greeks considered everything good in the world originated in Greece. Those who chose to remain in America became patriotic Americans, but they never lost their love and devotion to their native land.

Their native foods were important to them, and they included anise-flavored wine, black bread, fish, dried fruits, goat cheese, lamb and native confections.[106]

Italians

One of the two largest nationalities in the camp was the Italians. The Italians were willing to perform any job, either in the mine or outside. They were willing to work underground and/or pick usable pieces of coal out of the slack (waste) heaps outside the mine to sell. Some of the miners also had food wagons and sold popular Italian food after working a full shift during the day.

The Italians were well respected by their fellow immigrants/migrants. They were, for the most part, devout Roman Catholics and attended Holy Mass when it was offered.

Numerous married Italian women boarded unmarried Italian miners and also did washing and ironing to make extra money for their families.[107] Bill Deromedi's mother was one of those women. In the early years, the ladies used a scrub board and Fels Naptha soap (bar soap), which liquidized in a

The gas-powered washing machine was a lifesaver to the women who did many loads of laundry in a week. *John Holm, photograph by Levi Shinkle.*

pan on the stove. The clean laundry was then rinsed in the washer tub and hand fed through the wringer to get most of the water out.

In the late 1920s, the Maytag and Coronado companies manufactured gasoline-powered washing machines. They could be purchased at the Gamble's store in Thermopolis. The machines were single-cylinder and, later, twin-cylinder, which both created quite a racket as they agitated the water, soap and clothes in the tub. Each machine had a flexible exhaust hose that had to be vented outdoors through an open door. Many of the machines were also used on a porch, even in the winter. They were a huge help to the women who did laundry often every day.[108]

Scots

Most often, Scottish miners came from other parts of America. They were very willing to mix and intermarry with other nationalities. Iowa was one of the eastern states the Scots left to come to Crosby, Gebo and Kirby. They were drawn by the excellent reputation of the mining practices in place and of the coal itself. The religion practiced most often by the Scots was Presbyterian, but they were willing to attend other Protestant churches as well.[109]

Ruth (Johnstone) Woolman and Vernal Fred Woolman met and fell in love in Gebo. They each wrote a remembrance about her family's early days (and his dealings with the Johnstones) in Gebo, from 1923 to 1937. The remembrances have been combined here.

The Johnstones moved to Gebo from Iowa after selling their home there for $600. Edward (who was Scottish) and Bertha (who was Pennsylvania Dutch and an "amazing" cook) had a total of eight children: William and Katherine (neither lived in Gebo), Beulah, Alice, Mabel, George, Ruth and Robert (who all grew up in Gebo). Mrs. Johnstone "shipped a lot of the furniture to Gebo: her Edison Phonograph, the Singer Sewing Machine, the dining room furniture, beds, the rocking chairs that were the living room furniture."[110]

The home the Johnstones moved into had been Vernal's family home earlier. His parents had improved the house by "taking out a partition to make a living-dining room, and moving the windows together in that room." Vernal showed up at the Johnstones' front door shortly after they arrived, asking if he could "look in the cubby holes (that were under the eaves upstairs)" for his BB gun. It wasn't there. His parents had to get rid of it because he "went into the coal shed and shot at the Lynch kids through a knot hole." He was about seven years old. He remembered himself as "an ornery little kid in those days."[111]

After the Johnstones moved to Gebo, they got electricity for the first time:

> *They had strung electric wires into the houses and dropped a bare wire with a socket in the middle of the ceiling in each room. It took people awhile to realize they could get electric appliances. At first a light bulb screwed into the sockets with a string hanging to turn the light on or off was all you'd see. Then gradually the women found lamp shades for them. When Mama eventually got an electric iron she had to screw a plug into one of these ceiling sockets to attach to a cord, and position her ironing board below it to use the iron. After awhile she got a Maytag Washer. The clothes were put through a wringer, two revolving rubber cylinders, into a tub of rinse water on a bench next to the washer. Unluckily, the ladies* [sic] *long hair occasionally got caught up with the clothes going into the wringer. Ouch!*[112]

Vernal remembered when he fell in love with Ruth:

> *In the winter of 1933/34 on a snowy evening the kids were all out with their sleds, sliding down a gentle slope. Ruth had borrowed Sie Ashley's big Flexible Flyer sled, the biggest sled in town, and went up to get her pal Bunny to go sleigh-riding with her, but Bunny wasn't home. Lucy, her sister, volunteered to come along. As they were walking up the hill, they caught up with me, and Ruth asked me if I wanted to pull the sled up the hill. I pulled it up and then I rode down with them, several times. It was the start of a beautiful love. By the time I went home, I was already in love. My parents were in bed when I got home, and I went in and told them I had met the girl I was going to marry. I was 16 and Ruth was almost 14 at the time. Her mother had died a year before. We married on April 25, 1937.*[113]

Finns

Minnesota was originally home to Finnish miners who came to the district. They were hard workers and endeavored to stay employed year round by taking on non-mining jobs when the mining season ended each year.[114]

One of the Finnish miners was Eino Johannes "John" Pietila. He emigrated from Finland to Minnesota and then moved his family to Gebo

The early Eino Pietila family truly appreciated the life they led in Gebo. *Left to right*: Eino Sr., Eino Jr., Niilo and Sanni. Later, daughters Aira June and Maija Lilja joined the family. *Harri Joensuu.*

in the 1920s. He worked at Mine #2 until OCCC closed in 1938, when he moved his family to Washington State.

A cave that John made his own was discovered outside Gebo by Mary Gordon. He built a wall on the south end and put in a stovepipe heater. It could have been a "hunting hideout or man-cave or both." He identified the cave as "his" by etching his name and the year above the opening. Mary was "able to locate several of John's descendants and exchanged photos with, and heard family stories from, them. 'It's almost like he's part of my family, even though I'm not related to him or have ever met him,'" Mary stated. One of the photos included this statement about John's sons: 'Niilo and Eino like their life here. They have so much freedom here.—Gebo. Box 86 Wyo.'"[115]

Collectively, the Finns were known as superb stonemasons. Many of the finest stone structures in Thermopolis were crafted by Finns.[116]

Serbian-Montenegrins

The other largest nationality was the Serbian-Montenegrins. The term "Serbian-Montenegrin" was used to describe several sub-nationalities: Dalmatians (from Dalmatia, seagoing sailors), Croatians, Slovenians (both farmers) and Montenegrins (who were known as enemies of the Turks). Who were the "superior" Slavs? There were ongoing arguments.[117]

One of the biggest social events for the Serbians was a wedding. An open invitation was issued by the host of the celebration, and the festivities often went on for days. There was much cooking that went into the celebration: a whole roast lamb and pig that were cooked over an open pit; roast beef, ham and other meats; salads of many types, along with old country pickles and relishes; and so many desserts, such as apple strudel, pies and special walnut cakes that used ground walnuts to thicken the batter.[118]

After the feast, dancing commenced. Music was provided by a tambura orchestra. As the orchestra played, many folk dances were enjoyed. If a gentleman wished to dance with the bride, he was required to pin money to her wedding gown. Donald L. "Duke" Bolich, DDS, was born in Gebo, and his father worked in the mines for many years. When Duke and Sallie Bolich were married in 1957, Sallie's gown was covered with $1,200 in cash after the reception. It was understood that the money was given to "help the young couple get a good start."[119]

Punisha (Pete) and Kristinija Jurovich are shown on their wedding day in 1925. Bozo Jovanovic stands behind the couple.

The popular Montenegrin Tambura Orchestra is pictured in 1927–28. *Left to right*: Jerry Parazin, John Bolich, Mike Maravich and George Vucic. Seated is Mike Trbovich.

The Serbian-Montenegrin women were extremely hardworking. They often kept several boarders and did the washing and ironing for them. In addition, they took in laundry for bachelor miners. In the early years, the women had to use a washboard and washtub to get the laundry spotless.[120] Because they did so much cooking and baking, many of them had cook houses as well in back of the main house.[121]

Many Slavic men came to America when they were still boys of twelve or thirteen. They came knowing no one, except the relative who might have met them in New York City. Their priority was getting a job, even for little pay, and sending as much as they could back to the old country to prevent their families from starving.[122]

When a bachelor finally decided to marry, it was imperative that he stay within his native circle. The majority of these marriages were arranged by the men's families.[123]

The unmarried men who lived with families were eager to treat the children as their own. They even helped the boys attend college or start their own businesses. While Duke Bolich was attending college and dental school, he worked in the oilfields during the summers to pay for his schooling. Before his last year of dental school, he planned on working for a year to earn his tuition and fees. However, four bachelor miners arrived at the Bolich home, and each man presented Duke's parents with $1,000. "He must finish THIS year!" they all insisted.[124] The Serbian-Montenegrin girls were not encouraged to further their educations but to find acceptable, eligible men and get married.[125]

Irish

Many of the Irish miners came to the area from other parts of America, looking for a better working situation. It was easy to get along with them, and they were known for their senses of humor. They were historically devout Roman Catholics, and they normally looked for Catholic girls when they were ready to get married.[126]

English

The majority of English miners came west from eastern states, but some did come directly from England. They enjoyed the freedom they experienced in "New Amsterdam," as opposed to "the dull Puritan English colonies of their ancestors." Games such as dice, bowling and backgammon were commonly played and enjoyed.[127]

Many of the English won administrative positions in the mines because so many of them were literate. They could speak and read the English language easily.[128]

Welsh

Many Welsh men, both in Wales and in America, were historically miners, following in their forefathers' footsteps. They were well respected and easily intermarried with other nationalities.[129]

The Carey family came from Wales and then made many contributions to the camp. Their son John, who became known as "Count," learned many life lessons in Gebo, including the importance of participation in Boy Scouting and involvement in the community:

John "Count" Carey learned many lifelong skills while growing up in Gebo. As an adult, he became a Boy Scout leader and a Red Cross training volunteer.

A man by the name of Stanley Lynch, Sr. endeavored to start the first Boy Scout Troop in camp, and I remember some of the problems he had. It was all definitely military. We went there for military drill and we learned all of the different military maneuvers.

With the help of Rev. Hampton of the Baptist Church, several years later the Gebo troop did become quite well known. Someone in Crosby loaned us an old White bus that had a right hand drive with the gearshift on the outside and we toured all over Wyoming playing baseball games with different Boy Scout troops. We had a lot of fun.[130]

Count became a Boy Scout leader and Red Cross training volunteer in later years.[131]

Turks

There were a few Turks in the area. They had come to the United States looking for great wealth and then planned to return to their homeland. It was important for them to be able to eat their native food. They also proudly donned their native dress on occasion.[132]

Japanese

The area was home to one Japanese family, and the male members all worked in the mine. There was little discrimination against them, but the parents did not socialize with others.[133]

There were about twenty Japanese bachelors who lived in their own boardinghouse in Gebo. There was resentment against them because the mine brought them in to shovel coal into the railroad cars. Also, they were considered poor miners by many of the others.[134]

They were in Gebo for a relatively short time. An accident in the mine killed three of them, and the others all left the next day.[135]

American Common Stock

The majority of area residents were Americans. Most of them depended on the mines for their entire living. Others, however, mined in the winter and farmed in the summer. They had mined in other areas of the United States, but the West held some type of mystique for them.[136]

Swedes

The Swedes came to the area from other parts of Wyoming and America. In fact, many of them came after the only anthracite mine in Cambria, Wyoming, closed in the 1920s.

They gladly accepted American ways. Like most Scandinavians, they could speak at least some English.[137]

Poles

The Polish immigrants had come to America at a young age. They, too, planned to become wealthy and then return to their homeland. They were literate and could easily relate past history.[138]

When they emigrated from Poland, conditions there were extremely challenging. The aristocrats, and even the Roman Catholic clergy, held huge areas of land; but the poorer class had few chances for a decent life. They tried to make a meager living by farming small plots.[139]

When an aristocrat passed away, she or he was given a funeral Mass, sprinkled with holy water and taken to the cemetery in a procession. When a peasant died, they were not allowed a funeral Mass. The body was not even allowed inside the church. The priest sprinkled holy water on the body outside the front door of the church, and then the family and friends were responsible for interring the remains.[140]

Chinese

The Chinese were not included in Iris Guynn's original paper. According to Gebo resident Gary Giannino, the Chinese did laundry for the miners in the early days. They lived in structures called "hootches," or caves, and heated the caves with coals on the floor that they covered with some type of fireproof material. The heat rose and kept the caves comfortable. Eventually, they worked in the mines.[141]

In later years, Mileva Maravic, who graduated from Gebo High School in 1930, stated there were "very few Chinese and Japanese and the talk is they give them the most dangerous places in the mine. A cave-in or fall of rocks kills one and not very long after another Chinese goes the same way. Opinions soon become quite evident as to meaning in a small town." She donated her personal papers and memorabilia collection from Gebo to the University of Wyoming American Heritage Center.[142]

Duke Bolich had different memories of "discrimination" from his childhood. They were included in his autobiography:

> *It was a good place to grow up and know other nationalities. There was not discrimination in Gebo because almost all of the residents were coal miner's* [sic] *families and basically poor. One of the nicest things about living there was being invited to dinner by so many mothers making dinner. We often had some of our friends over for dinner with us, too. We experienced so many kinds of ethnic foods and enjoyed all of them.*[143]

Buildings and Businesses

The center of the camp was the gigantic Mine #1 tipple. The mining operation also required a number of additional structures, including the hoist house; the blacksmith shop (the coal in Gebo was of such high quality that the blacksmiths from the Hanna, Wyoming mines came up to Gebo to

The tipple was the most noticeable component of the complex Mine #1 operation.

The Mine #1 blacksmiths were Eli Talovich (*left*) and Ray Dickey (*right*). In their shop, they built and maintained the coal cars.

buy coal); the fans (which provided the crucial ventilation to the mine); the shower house; the spray ponds and engine house; and the concrete bins at the bottom of the tipple (stored the coal from the tipple).[144]

With hundreds of employees necessary to run the mine, and eventually Mine #2, a large company building was built by OCCC across from the tipple.[145] The mine office was located on the Gebo version of "Main Street." There were no sidewalks or gravel streets, just dirt roads.[146]

The bookkeeping was done in the office, along with maintaining the miners' time sheets. Every two weeks, the miners were paid in cash, and the money arrived on the train from the bank in Thermopolis. Deductions were made from each salary: hospital charges, union dues, road and school taxes. The taxes did not amount to a significant sum, but the miners resented them. They were particularly despised by the foreign-born miners.[147]

A grocery-type store, the company store, was built by the OCCC. It was managed by Robert C. Rae. Even though the quality of the available merchandise was good, there were complaints. Most items were more

The "Main Street" of Gebo was home to businesses and homes as well. *Mileva Maravic_6309_Box1_Folder Gebo Town_MainStreet (2), American Heritage Center, University of Wyoming.*

Robert C. Rae was the manager of the Owl Creek Coal Company General Store in 1914. *Mileva Maravic_6309_Box1_Folder Gebo Residents_1914–1918, American Heritage Center, University of Wyoming.*

expensive than what was available at the privately owned stores in Kirby. Nevertheless, the customers were able to order items each day and have them delivered to their homes the next day.[148]

Another advantage of having the company store was that the customers could receive a coupon book. Each book was worth ten dollars, and the coupons were used to buy various items. On payday, the store had to be repaid for any credit that had been extended to a miner.[149] There was a joke in the camp about the miners' wives on payday. The women normally didn't want to "get close" to their husbands until a proper shower/bath had been taken at the end of the workday—except on payday. Then, wives' fingers were going in overall pockets to make sure that the money went safely into the "proper" hands.[150] The post office was also located within the store. The bank building would have been visited on payday as well.[151]

The meat market was separate from the grocery store, located in a wooden frame building. The cuts of meat were displayed where they could be easily seen, and the customers watched the butcher slice the meat to order. The customers watched the scale to ensure that the amount charged by the butcher was correct. Customers could also request that the meat be cut in a certain way, as the cattle were purchased by the butcher and slaughtered to satisfy the customers.[152]

When automobiles came to Gebo, a garage building became necessary. The company obliged by building one. Very few people were able to afford such an extravagance. But those who could were certainly envied.

The pool hall was a very popular gathering place for the men. *Mileva Maravic_6309_Box 1_Folder Gebo Town, American Heritage Center, University of Wyoming.*

Model T Fords required a great deal of care, and their owners were able to rent personal garage space. Many hours were spent keeping these prized possessions in tip-top shape.[153]

The vehicles needed fuel and oil, so a filling station came next. The owners could watch as the gas was hand pumped, and they knew exactly how much gas their vehicle required. The oil was kept in refillable quart bottles. A pouring funnel was simply attached to the bottle, and the station attendant filled the oil pan to the proper level.[154]

The men wanted a place to gather, and that was the pool hall. Many activities were enjoyed there: "playing pool and cards, and buying liquor, ice cream and candy. On the nights the hall was open for dances, food such as sandwiches and chili were available for purchase."[155] Mileva remembered that children were not allowed in the pool hall itself. They were required to enter by the back door. There they waited until the clerk came to ask what kind of candy bars and/or soda pop they wanted. If they didn't have cash with them, they could buy items with pool hall chips miners occasionally gave them.[156] The pool hall was managed by Bob Rae.[157]

A large community building was the scene of many social events. It was the center for numerous enjoyable times shared by Gebo residents. Because there were few opportunities for outside entertainment, "good times" came from the residents themselves. The numerous nationalities provided opportunities to showcase talents. Amateur nights were very popular, as anyone could perform. Native songs, as well as American patriotic tunes, were heard. Poetry recitation was certainly appreciated. There was much excitement when a play was planned, as there were numerous roles that needed to be filled. Joining together to sing along with the piano encouraged a feeling of camaraderie among the singers.[158]

Lodges were essential to many groups in Gebo, particularly immigrants. The lodges provided "a common brotherhood, many times throughout the U.S." The lodges offered added benefits, including insurance, which gave the members a feeling of extra security. The men often remained active for their entire lifetimes.[159]

A union hall was built by the United Mine Workers of America, Union #2671. These meetings were well attended. Mandatory monthly dues were paid, and there was little resistance to this necessity. The miners recognized the advantages of belonging to the organization, including the bargaining power of the local and national groups regarding salaries and safety concerns, among other topics. Again, many miners boasted lifetime membership. Their committal services were conducted by the unions as well.[160]

Getting a professional haircut or having a beautician style one's hair required a trip elsewhere, until Gebo was able to attract a barbershop (a small building near the front of the pool hall; it was off-limits to women)[161] and a beauty shop. Not everyone was able to afford such luxuries, mostly bachelors and those married miners with the higher-paying positions.[162]

The OCCC had rooms available to rent, and many of their employees took advantage of this situation. The building was called the Owl Creek Boarding House. Many of the bachelor miners required a place to sleep and meals for a reasonable price. Approximately 120 miners ate breakfast and dinner and slept at the boardinghouse. Many women were employed to keep the meals prepared and served, as well as to keep the sleeping quarters made up and cleaned. They made a handsome salary for that time, ninety dollars per month.[163] There were privately owned boardinghouses as well, run by Mrs. Budna, Mrs. Roganovich and her daughter Mary and Mrs. Whaley.

The few Japanese miners in Gebo had their own boardinghouse. They had the luxury of enjoying their home cuisine, and the cook was also their housekeeper. The house had the reputation for being spotless. Rather than being designed like a typical American boardinghouse, this one had the bedrooms on the main floor and the dining room on the second floor.[164]

The Finns also had their own boardinghouse, which cost five dollars more per month than the Owl Creek Boarding House. They also enjoyed their home country's cooking, which they considered better than any other. So the extra money they had to pay was "well worth it." An integral part of the Finnish boardinghouse was the sauna. Enjoying a warm, soothing sauna bath was well worth the extra fee charged.[165]

One of the reasons so many men wanted to work for Owl Creek Coal Company was its reputation for being a safe place to work. In 1886, Wyoming territorial legislators passed new mine safety laws:

> *Similar bills had been introduced in earlier legislatures, but no action had been taken. The new law created the office of state mine inspector with the duty of inspecting every coal mine in Wyoming no less frequently than every three months. Further, the act banned boys under 14 from working underground. Women were also "protected" from the hazards of mining when it was made unlawful for them to work in mines.*
>
> *In 1889, when Wyoming was drawing up a state constitution, delegates adopted much of the territorial law as Article 9 of the new constitution. Included was the provision: "The legislature shall provide*

> *by law for proper development, ventilation, drainage, and operation of all mines in this state." The delegates endorsed both the earlier territorial law giving women the right to vote and the earlier ban on their working in mines.*[166]

Many measures were continually employed to prevent accidents from occurring. One later measure stated that miners were forbidden to smoke in the mines. Another included forming a Mine Safety Team (which later became the Mine Rescue and First Aid Team). The ability to rescue miners in distress and treat their injuries in the mines was imperative. The OCCC constructed a building to house the team. The members practiced first aid procedures as well as other necessary skills. There were statewide competitions to showcase the skills of various mines' teams, and the Gebo teams won many awards. They even won third place at the international contest one year.[167] The early team carried a caged canary in competition, which was used in the mines to alert the miners about bad air conditions. Their equipment also included gas masks, a stretcher and safety goggles. Tony Pisto directed the early team, and Telford McCallum and Donald McCallum were members of the later team.[168]

Many of the miners were highly musically talented, and a Gebo Miners Band was formed. The members were from all nationalities. The band gained a reputation as a top-notch organization and performed at

The early Mine Safety Team presented a demonstration for the hometown crowd in the Thermopolis gymnasium. Tony Pisto is on the far right.

The Gebo Miners Band was well known far and wide. Stephen A. Sable, president, and Bob Rae, director, are pictured in the center.

numerous venues, including the Wyoming State Fair in Douglas. They entered numerous competitions and won first place often, which brought a real sense of pride to the Gebo residents. Robert C. Rae was the director. The group's practices were held in the Community Hall, and the band performed year round. They wore winter and summer uniforms; their summer attire included their mining caps. The concerts held in Gebo were always well attended. The OCCC carried on its tradition of benevolence by building an outdoor gazebo near the pool hall that was used as a warm-weather venue for the Gebo Miners Band.[169]

In the 1910s, a Women's Band was also formed. It was basically a drum and bugle corps, with brass instruments and drums borrowed from the miners' group. The women and girls also performed marching maneuvers during the warm-weather months. As women never considered wearing pants in public, the band performed while wearing ankle-length dresses or skirts and blouses, dress shoes and large dress hats. The tuba player was not required to wear a hat. Occasionally, a few brass players from the Gebo Miners Band joined the ladies. They rehearsed and performed in the same

Top: A Gebo Women's Band was formed in the 1910s and 1920s. They were basically a drum and bugle corps, with brass instruments and drums.

Bottom: The first houses in Gebo were built by OCCC and were never meant to be permanent. They featured identical design and inferior building materials.

locations as the Miners Band. They were considered excellent musicians and were warmly received. The band was in existence only during the 1910s and 1920s.[170]

Sam Gebo had envisioned a "model city" in the Kirby area, with companion industries to grow up to make "year- round work for the men." The houses that were first provided for the Gebo miners were designed to be temporary and were poorly built, as it was thought that everyone would eventually move to Kirby.[171]

When it became evident that Sam's vision for Kirby would not materialize, the company provided more substantial homes. They were larger but not necessarily better built. Each had a small hut behind the main

Milovan Jovanovic and his sons Bunky (*center*) and Richard (*right*) are shown in their yard in "Silk Stocking" or "Rock Row."

house, which was used to keep the coal and wood dry for the coal stoves that everyone depended on. The homes had four rooms on the lower floor: a living room, dining room area, kitchen and bedroom. Upstairs there were two bedrooms. There were also outhouses (outdoor toilets) that were placed away from the homes. The houses had no insulation in the walls or ceiling and were terribly drafty and cold during the winters. A common joke was that "you could throw a cat through the walls." Everyone had to gather around the coal stove in the corner of the living room to keep warm. The kitchen stove kept the kitchen and dining room warm. There was usually one small vent above the living room stove that gave a little (but not much) heat to the upstairs bedrooms. All of the windows leaked a lot of air because they weren't insulated either. When the wind blew, the kids used to say, "We heard the windows hum." Many people had drapes to close after dark to help keep the rooms warmer.[172]

Some miners were financially able to build their own homes. They owned the buildings themselves, but the land still belonged to the OCCC. One of the "exclusive" sections of Gebo with miner-owned homes was called Rock Row (because the walls were made of stone) or Silk Stocking Row (because the ladies who lived in these homes could afford to buy silk stockings).

MEDICAL CARE

A twenty-room hospital was built and maintained by the company, and the "contract surgeon" and nurses were employed by Owl Creek Coal Company. The company was also responsible for providing a home for the doctor and one for the nurse. The residences were built next to the hospital, on a hill southwest of the tipple. The miners were charged one dollar per month to cover any medical expenses they incurred. Eventually, that figure rose to three dollars per month.[173]

According to Hot Springs County historian Ray Shaffer, a diphtheria epidemic went through the district in the early 1920s, with "the highest death toll occurring in 1921." There was a shortage of diphtheria vaccine, so each household received only one dose. The airborne virus was highly contagious and particularly dangerous to children. The victim's throat swelled up and choked off their ability to breathe. Suffocation followed. Parents faced a heart-wrenching decision: if they had more than one child, who got the precious dose? In many families, one child survived and

Gebo also had a beloved doctor, Dr. J.R. Pierce Sr. He served as Owl Creek Coal Company physician for many years. His practice also included Thermopolis, where he lived.

the rest perished. This terrible time was in addition to the 1918 influenza epidemic that struck the entire world.[174]

There were a number of physicians over the years who staffed the hospital, including Dr. Williams, Dr. G.M. Penfold, Dr. C.C. Hickman and Dr. J.R. Pierce (who also practiced in Thermopolis for many years). After Dr. Pierce moved into Thermopolis, he still cared for the Geboites. The story was that he "drove his Pierce Arrow 100 miles per hour on an emergency" getting out to Gebo. When Dr. Pierce left Thermopolis, he was replaced by Dr. Tennicke.[175]

The doctors, especially, were admired throughout the community. The level of care they provided was greatly appreciated. There was a story about one doctor who played cards with the miners and "even went outside the pool hall to roll dice on the smooth ground."[176]

During the time the miners were still working six-day workweeks, they lived in constant darkness. Duke Bolich remembered:

> *In the winter months, dad would leave the house about 7 a.m. and walk up to catch a bus to take him to the Miller Mine where he worked at that time. It was a few miles to the west, but too far to walk. He would work all day in the mine and then be home shortly after 5 p.m. In the winter, the sunrise is close to 7:30 and sunset around 4:30, so he never saw the sun until Sunday. It is no wonder that he suffered from depression.*[177]

Schools

All of the school grades were housed in one building. As the population of the camp increased, more grades and rooms were added. The high school was the last to be built, on the right side of the building. The Wyoming Board of Education "rated the high school as excellent." This distinction gave the school much prestige. The middle portion of the building housed the fifth through eighth grade class area. To the left was the primary school area (grades kindergarten through fourth grade).[178]

Gebo School was a source of community pride and the scene of countless community activities.

The members of a Gebo School seventh-grade class, and a little "visitor" in the middle row, are pictured in front of their school building. *Mileva Maravic_6309_Box1_Folder Gebo Schools-Students and Teachers_1915–1930, American Heritage Center, University of Wyoming.*

The 1930 graduating class of Gebo High School. Mileva Maravic stands in the back row, right side. *Mileva Maravic_6309_Box1_Folder Gebo Schools-Students and Teachers_1915–1930, American Heritage Center, University of Wyoming.*

Gebo School's 1936 third- or fourth-grade class is shown with their teacher, Miss Mary Kay Burnell.

The high school colors were gold and maroon, and their nickname was the Miners. A cheer heard at sporting events was:

Baby in the high chair,
Zis Boom Bah!
Gebo High School!
Rah! Rah! Rah![179]

Mileva Maravic added, "Away from the playground and the school were two small buildings near a pile of large rocks. They were the outdoor toilets. One marked BOYS and the other GIRLS."[180]

Mileva also recalled the academics and what the girls wore to school in the winter and early spring:

> *The Owl Creek Coal Company furnished the books. The children bought paper, pencils, penholder made of wood into which you inserted a metal penpoint. We also had to buy a bottle of ink which cost ten cents. The desks had a round hole in the upper right hand corner for the bottle of ink. The penholder with penpoint dipped in the ink we used when having Penmanship Class. We practiced many days the push-pull exercises and making large round circles before the teacher accepted the papers which she sent to the Palmer Method Company in Chicago. They graded the papers and if satisfactory were returned to the School with a Certificate having our name on it. We also received a Palmer Method Pin we could wear.*
>
> *In winter we wore long underwear, heavy brown or black hose that came over the knees, high button shoes, later laced shoes. If there was a rash of colds or flu—Mother would put garlic cloves in our pockets. If we got a cold, we ate lot* [sic] *of garlic at home. According to researchers at George Washington University eating onions and garlic is good for one's health if the onions or garlic are eaten raw, cooked or taken in extract or tablet form but Dr. Vanderhock says they won't work if you just hang them around your neck.* Billings Gazette *newspaper, December 1979. We didn't know that.*
>
> *In early Spring still the long underwear—on the way to school I would stop in someone's outdoor toilet, roll the underwear from ankle to above the knee. After school on the way home stop again to roll it down.*[181]

The students had a merry-go-round and a very tall metal slide to play on. The whole camp was very proud of the school. The company paid

School field trips were shorter and simpler in Gebo in 1915. The students simply hiked to one of the many rock outcroppings around and outside the camp. Miss Ellen Porter and Miss Baer are pictured with the students.

the teachers a decent wage. The rooms all had a good-sized coal stove to keep the rooms warm in the wintertime. Everyone had to neatly place their overshoes and boots in the foyer so that they did not bring mud and snow into the classroom.[182]

Even though there were few opportunities to travel farther than to Thermopolis, the students still had the chance to take field trips. One of the favorite getaways was to the various rock formations found in the Gebo area.[183]

For the parents who hadn't had the opportunity to receive an education in their native lands, seeing their children educated brought much parental pride. Some of the fathers and other miners were able to realize a dream: to obtain a basic education themselves. The company arranged for them to attend night school and learn to read, write and do basic math. Possessing these skills gave the men a real advantage in everyday life. They attended classes at the school, with the understanding that there was to be no smoking

or spitting on the floor. The school board insisted that this rule be strictly enforced. No women were allowed to attend the classes.[184]

In late 1912 and early 1913, the paymaster of OCCC decided to teach writing to some of the men. This presented a win-win situation: the men became more literate, and they were able to assist the paymaster in his duties.[185]

Faith

With the numerous nationalities in Gebo, there were numerous religions practiced as well. According to T.A.L. Nicholas, almost all denominations shared the community church. Several itinerant preachers, like Reverend Kelly from Billings, came through Gebo to preach to the faithful. Reverend Clyde Hampton oversaw a very large Sunday school class.[186]

Father Endres from the Catholic church in Thermopolis drove out to Gebo one or two Sundays a month. He arrived at 5:30 a.m. to offer Mass at the Union Hall. During the service, his team of horses was fed and watered by miner volunteers. After Mass, he was given breakfast by one of the parishioners. Then, he was on his way back to Thermopolis by 7:00 to make the three-hour trip for Sunday Mass in town. His travels were often made through snow and wind.[187]

Camp "Services"

Because Gebo was never incorporated, the Geboites were not entitled to traditional municipal amenities. However, industrious citizens took it upon themselves to provide necessary services. In the early days, before water was piped to Gebo from Kirby, the "water man" delivered water from a wagon every day:

> *This semi-arid land was located about five miles from the Big Horn River and drinking water. Consequently, the water had to be hauled in by horse and wagon with a large wooden drum serving as the tank. The individual who hauled the water obtained it from the Big Horn River and laboriously filled the tank, bucket by bucket, until the tank was loaded to the brim. At each house and at the various business houses there were barrels to be filled for a fee of 25 cents per barrel.*

The clean white laundry was proudly shown off by Gebo's women.

At that time much pride was taken by the women to have the cleanest clothes on any clothesline as well as being the first to have their washing completed and hung upon the line. To take advantage of every opportunity, some of the eager women would wait for the water wagon in the narrows and bribe the individual who drove the wagon. Spirits made at home were the usual briberies, and if the driver imbibed too much at an early hour it could easily be the end of the water for that day. This caused much animosity because the water was needed for drinking as well as for cleaning purposes.[188]

In 1913, the district became part of the newly formed Hot Springs County. Mine superintendent Barnard was elected as one of the first county commissioners. Even though Gebo had not been previously incorporated, it still required a law enforcement and judicial presence. Barnard was able to have Dick Kirby appointed constable and Jim Rae appointed justice of the peace.[189]

Both were necessary, as a number of bootleggers had set up shop in the area. They eventually came up with an effective plan to stay one step ahead of the law. They dug under the numerous rock outcroppings in the area and made small "rooms" called blind pigs. They were reinforced with stolen railroad boxcar doors and tar paper. The term "blind pig" came from the

state of Maine. A tavern owner sold tickets to his customers that entitled them to see his blind pig. Once the customers were inside, the tavern owner gave them a free drink of bootlegged liquor.[190]

In the spring of 1913, it was determined that it was time to close down the blind pigs. A raiding party was formed in the OCCC office to serve subpoenas to every bootlegger. However, when the Hot Springs County deputies entered the various rooms, the bootleggers were not to be found. Being miners themselves, the bootleggers had dug escape tunnels.[191]

It was assumed that at least some of the bootleggers could be caught and subpoenaed. However, when daylight broke, it became apparent that wasn't the case. The bootleggers came into view, running across the hills. They were being pursued by the deputies, who were shooting into the air and hollering, "Halt!"[192]

Eventually, several of the bootleggers did return to the Gebo area.[193]

The excerpt below is from Mileva Maravic regarding "The Roaring Twenties and Prohibition in Gebo":

> *At age fourteen I knew how wine was made and how to cook whiskey. I loved September. September meant a new year at school. It also meant we would have all the fresh grapes to eat when they came from California. In August one of the miners went around camp taking orders. How many boxes of grapes and what kind red or white did they want? My step-dad ordered many boxes mostly the red, some white grapes for the white wine. I remember people talking about ordering one ton, half a ton of grapes. It took many boxes to fill the fifty gallon wooden barrels. Several boxcars of grapes came to Kirby from California. The grapes were brought to Gebo by truck.*
>
> *When it was wine making time my step-dad* [Eli "Smokey" Talovich] *ordered us kids not to bring any of our friends home from school. They may see the wooden barrels, the grapes, or smell the odor coming from the dirt cellar under the house. It wasn't an easy thing to do to keep our friends away. The friends may tell their parents, who might report it to the Revenue Officers in Thermopolis. To me that was a bit puzzling as our friends had the same thing going on at their house.*
>
> *A funny story went around camp what a miner said to the Revenue Officers when they came to his house asking what he was going to do with all the grapes he had. He said: "the wife she is going to make a little jelly for the kids."*

When I left home for the University of Wyoming and heard whiskey was made from corn, wheat and potatoes, I was shocked [and] *thought how awful that stuff must be. I assumed all whiskey was made from grapes like we made it in Gebo. I learned it was Brandy we made when it came from grape mash.*[194]

GEBO'S WAR EFFORTS

In the spring of 1912, many Gebo miners were called back to their homelands for military service. In order to pay for their trip home, many of them dug up money they had hidden in the hills around the camp. Much of the money was wrapped in "oily bacon paper," which the miners had gotten at Mrs. Budna's boardinghouse.[195]

Twenty-three Serbian-Montenegrin men from Gebo volunteered to fight for America during World War I. They joined the U.S. Army and all left for basic training together.

As more and more European countries entered World War I, men from Gebo went back to their home countries to join the fight. These twenty-three Serbians stayed and fought for the United States, circa 1917.

Many farewell parties were held for all the men. Thomas A.L. Nicholas, who worked in the OCCC business office in the 1910s, told the story:

> *It might be in some miner's home and even out of doors in the snow. They would buy a lamb from some rancher and roast the lamb on a spit, or over a bonfire. You could see Greeks dressed in Evzone skirts, passing bottles of spirits and Greek wines scented with anise or licorice to Turks and Servians* [sic] *and Bulgarians, Montenegrins and Austrians with who they worked and played pitch and formed a warm comradeship. A day or two later a dozen of these men would mount Miller's stage and ride away together to take the train at Kirby, the first step on the journey. Soon they would join opposing European armies, and be shooting at one another in the Balkan War. Few of these men ever returned to Gebo.*[196]

The coal from the mines in Gebo and Crosby was vital to the war effort during World War I. However, it was difficult to maintain a consistent production schedule in the mines:

> *During December, 1917, we received complaints from the operators in Hot Springs County that the saloons at the coal mines were open day and night, demoralizing the labor force and causing a decrease in production. We requested Col. George M. Sliney, Fuel Administrator for Hot Springs County, to investigate and do everything possible to regulate the disturbing element. Colonel Sliney took vigorous action and regulations were passed by both the Civic and County Authorities closing the saloons from 11:00 P.M. to 7:00 A.M. and all day on Sundays. This had the desired effect for a time but it was necessary to continuously guard against the abuse of liquor causing a decrease in production. The mines in Hot Springs County are located in a bleak and desolate section, practically a desert, and opportunities for amusement and recreation are limited.*[197]

CEMETERY

The Gebo cemetery contains headstones with many "melting pot" surnames. The majority of the markers are tiny graves, along with baby crib and child-sized bedframes, remembering children who passed much too soon. The influenza epidemic of 1918 took its toll on all age groups.

The Gebo cemetery reflects the fact that mostly children rest there. *Photo by Levi Shinkle.*

As in Crosby, many children perished in the diphtheria epidemic in 1921. Most of the Gebo adults are buried in the two Thermopolis cemeteries.[198]

Unfortunately, many of the markers, as well as a small house-type structure, have been desecrated. A fence has been erected around the area.[199]

SPORTS

The OCCC installed a baseball diamond for the miners. The first team was made up of mostly Iowa miners, who had played together before coming west. The team's nickname was the "Three I League," but they were known as the Gebo Miners Team when they played against other towns. Because so many of the players had played together for years previously, they understood teamwork and exhibited excellent skills. They won many victories over the years.[200]

When he was older, Count Carey was able to play baseball with the Gebo Miners baseball team:

> *Another thing that Gebo taught me was the value of athletics. I think that this was due to several men like Dave and Bud King and Ricky Humme*

who were good ball players and encouraged we boys to play ball. Like me, I was 15 years old, and they would let me play for an inning or so.

I remember in particular one home run which Bud King hit. We later measured it as a scout project, it was over 450 feet in the air before it hit the ground.

Gebo had a really powerful ball club. Anyone that played ball and worked in the coal mine got top rating. My experience playing ball there enabled me to play nine years of Class C ball with Hamms Beer in Denver. It entitled me to two free trips for tryouts in the Southern Association at San Antonia [sic] *Tex. I didn't like it there, however, and wouldn't play ball so I got shipped back to Denver again, but it was still quite a thrill to have the honor of being picked to go there.*[201]

Duke Bolich also related the importance of baseball to the Gebo boys:

In the spring and summer, we would play many sports. If it was baseball, we would go over to the boy who had a bat and another boy who had a ball and then organize a game. We would use a flat rock for each base and home plate. I can still visualize the ball, which was wrapped in black friction tape, flying off of the bat with the tail of the tape trailing the ball.

The Gebo Miners baseball team is pictured in 1925–26. They experienced much success over the years. *August French.*

We did not have gloves or catcher's masks. If you were an infielder and the ball came to you, you tried to knock it down and then pick it up as it would sting your hands. We seldom had enough players to have a full team so we would rotate from batter to catcher, first base, second base, short stop, third base, left field, center field and right field. If we had 10 it was great for the rotation, but sometimes we had two outfielders and two infielders. Since I was one of the younger kids to play, I was always stuck out in the outfield where I hoped they would not hit it my way. I always hated the dirt that stuck to the friction tape that covered the ball. It was only when we moved to Thermopolis and played American Legion baseball that I really enjoyed a clean white ball devoid of sticky friction tape and dirt.[202]

According to Duke, basketball was also a big favorite:

Everyone loved to play basketball. We played in the room at the school in the winter. My brother Ken came up with the idea of ordering letters to sew on T shirts for our team. Everyone decided on using UMWA (United Mine Workers of America) since we were coal miner kids. Once they were sewed on, it was confusing because if we had a scrimmage, we all looked alike so you had to know who was on your team by their face and not by their uniform.

It was wonderful that we had such an interest in basketball. It paid off in future years because Gebo kids were the core of the Thermopolis High School teams. In my high school years, Ken Bolich, Jack Bretthauer, Tritz Jurovich, Jack Toth, Milo Galovich and George Todorovich (Kirby) were members of the team along with Moe Radovich who was born in Crosby, two miles away, but grew up in Thermopolis. Moe later became an All-American basketball player at the University of Wyoming in 1952. John Pilch was from Gebo originally before moving to Thermopolis. His father was the director of the Gebo Miners Band in the 1920's. He was selected on the 1950 All American basketball team. It is so unusual to have two All American basketball players from the University of Wyoming who came from two ghost towns in the same county![203]

A favorite basketball locker room story at Thermopolis High School reflected the coal mining influence on the circa 1940 team. The Thermopolis Bobcats played a game at Casper High School. The announcer was reading the Bobcats roster over the loudspeaker: "Galovich, Ranovich, Todorovich, Jukanovich, Raicevich, Jurovich, Tsubovich and Koebelin." The announcer "blurted out," "Koebelin? Now how the hell did he get in there?"[204]

Left: Moe Radovich grew up in Crosby and played for the Northwest Center, the University of Wyoming and in the NBA. He also coached the Wyoming Cowboys from 1974 to 1976. *University of Wyoming Sports Hall of Fame.*

Right: Gebo's John Pilch earned All-American honors at the University of Wyoming and then played in the NBA. *University of Wyoming Sports Hall of Fame.*

Tennis was a popular sport in Gebo as well, and a court was built for playing and watching. The younger set, and store employees, were the ones who most frequently played. The court was even lit at night by large-watt bulbs, which allowed for evening matches and brought a sense of pride to the players.[205]

Entertainment and Fun

Thermopolis, which was twelve miles away, offered many forms of entertainment and fun. The biggest attraction was the state park, which was advertised as "Home of the World's Largest Mineral Hot Springs." After being underground for eight hours during a workday, many miners developed arthritis and rheumatism (any disease marked by inflammation and pain in joints, muscles or fibers). They found relief in the warm mineral waters of the commercial swimming pools and the free state bathhouse. Also, being out in the open gave a feeling of freedom from the tensions of being in the dark underground, and the sunshine was helpful in returning some color to the miners' pale skin.[206]

Besides enjoying the soaking and swimming opportunities in the state park, visitors could also admire the travertine terraces that were prominent in the area. The minerals in the hot water formed beautiful colored patterns, and these were visible along the walkway that meandered around the hot springs. The water flowed from the Big Spring and dropped forty feet off the terrace edge to the Big Horn River below. Herds of buffalo and elk could also be seen, and many picnic areas were available for use.[207]

Thermopolis also boasted a movie theater and later a drive-in theater; numerous downtown clothing, hardware and gift stores; pharmacies; restaurants and bars; florists; jewelry stores; toy stores; grocery stores; banks; and other businesses.[208]

One Saturday, the Bolich family was invited to go to "town" (Thermopolis) with a neighbor who owned a car. Duke related the story:[209]

> *Mom sent Ken and me to the Tepee Movie Theater to attend the afternoon matinee. The admission for children under twelve was 11 cents, with a charge of 35 cents for adults.*
>
> *I do not remember the name of the main attraction, but I do remember watching "The Three Stooges" in a comedy short that preceded the main movie. One of the Stooges escorted a beautiful lady, who wore a long evening dress, to the table. One of the Stooges, named Curley, pulled out a chair for the lady to sit down. When she started to sit, he pulled the chair out from under her, and she hit the floor. Everyone in the theater roared with laughter. People were still laughing when the short movie ended.*
>
> *A short time later, a family friend came to visit my mother on a warm summer day. She always wore a house dress and had her hair in a braid, which she wrapped around her head so that it was not hanging down. She then placed a little black decorated hat on her head and held it in place with hair pins. As little boys, we thought that she was bald because of the hair decorations.*
>
> *My mother came in from the back porch and greeted her friend and asked her if she would like to sit down at the dining room table and have a cup of coffee. My mom pulled her chair out to sit down, and like a polite little boy, I pulled out her friend's chair from under the table. She turned to me and said it was so polite of me to pull the chair out for her and then proceeded*

to sit down. As she went to sit down, I pulled the chair toward me and sure enough, she hit the floor. She was an older woman, so we were lucky that she was not hurt. My mother grabbed me and gave me about five paddles.

Duke also discussed the camp's favorite activities:

Our favorite winter sport was in two areas. First, there was a hill near the Pool Hall that was called "Bumpety Bump." It consisted of a hill with three dips and three mounds, so when you went down the hill on your sled you would go bumpety bump.

In the spring of the year, the water tank on top of the hill a short distance from our home would freeze and overflow. The overflow would come down the hill fairly close to our home. We would get our sleds, walk up the hill to just below the water tank and then place our sled on the ice. It was a great ride down the hill when it was cold enough so the ice was not melting. Many times, it was very slushy so when you would crash your sled, you would get soaked. Since our house was the closest to the sled path, kids all came to our house to dry out.

We had a large coal stove in the corner of the living room. It was a perfect place to get warm and get dried out. My mother always had some of my dad's shirts handy to have kids get out of their wet clothes, put on a shirt and sit by the fire. She would get the ice cream chairs from the dining room, put them behind the stove, and put the clothes on them. It was fun to visit with our friends for 10 to 15 minutes while their clothes dried. Then, we were back on the hill again. Sometimes, we would have eight or ten friends around the fire.[210]

Ice skating was enjoyed by all ages, and Gebo had two rinks over the years. When the high school gym was torn down, the vacant area was flooded in the winter by the miners. A reservoir was built by the Department of Interior east of the camp, and it froze over every winter as well.[211]

Halloween was another holiday that kids enjoyed. We would go house to house and knock on the front door. When they opened the door we would yell, "Trick or Treat." The person answering the door would dole out a piece of candy to each one of us. Some of the older kids would soap windows on houses for Halloween. Others, who were probably high school age, would tip over the outhouses that people used for their toilet. The next day, we often saw the men getting together to upright the outhouses. As I grew a little older

> *and was out a little later, I would see some of the older kids tip over the outhouses. I don't think it lasted much later, because they really emphasized the trouble you would be in if you were caught. So after that, it was more soaping the windows which would require the people to clean the windows beginning November 1st.*[212]

Kirby was also a destination for more socializing, entertainment and fun. Choices there included buying groceries or going in the saloons, the hotel, the train station and the post office. A house of ill repute known as Gertie's was a popular choice, because there were so many unmarried men in the district. They could hear Rosie play the piano, buy a drink and secure the services of one of the "girls" for three dollars. It was well known that Gertie's girls had to be "checked" periodically by a local physician to make sure no venereal disease was present.[213]

CELEBRATIONS

There were two important dates that were celebrated each year by the UMWA, June 1 and Labor Day. The June 1 celebration was always hosted by Crosby and commemorated the day when an eight-hour workday for miners officially became accepted.[214] All ages were invited, and there were activities for everyone. Races of many types brought cheers from the enthusiastic onlookers. The children attempted to catch a greased pig, and climbing a greased pole challenged older competitors. Traditional sports, including boxing and baseball, were enjoyed by all.

There was food of every type offered, and many "old country" recipes were on display. Soft drinks for the children and moonshine for the adults were also served, and good fun was the order of the day. The merchants from Thermopolis made donations of various types, as their livelihoods certainly depended on the hardworking people at the festivities.

The big Labor Day celebration occurred on the first Monday of September in Gebo and always meant a delightful three-day weekend. Besides the fun-filled activities enjoyed on June 1, Labor Day also included time to visit with those miners and their families who had moved to other camps. Miners were known as a transient population; often, miners from the district moved to camps in Red Lodge, Bear Lodge or Butte, Montana; Hudson, Wyoming; or to one of the camps located on the Union Pacific line along the southern border of Wyoming.[215]

A bonus was the prize money awarded for the various contests. The prizes varied from fifty cents for the under-six-year-old races, to two and three dollars for the boys' smack-a-feller-game, to twenty-two dollars for the winning team in the tug-o-wars. There were races for children, in various age groups, for both boys and girls. For the men, there were numerous contests and races, including men's free-for-all, horseshoe pitching, fat man's race and men's sack race, among others.[216]

Women also had the opportunity to compete in additional contests: husband calling, coal carrying, balloon carrying and married lady foot

LABOR DAY CELEBRATION

Boys under six years	1st prize 50c		all others 25c
Girls " " "	" " 50c		" " 25c
Boys Race 6-8 "	" " 1.00	2nd–75c	3rd-50c
Girls " " "	" " 1.00	" 75c	" 50c
Boys " 8-10 "	" " 1.25	" 1.00	" 75c
Cirls " " "	" " 1.25	" 1.00	" 75c
Boys " 10-12 "	" " 1.50	" 1.25	" 1.00
Girls " " "	" " 1.50	" 1.25	" 1.00
Boys 12-14 "	" " 1.75	" 1.50	" 1.00
Girls " "	" " 1.75	" 1.50	" 1.00
Boys 14-16 "	" " 2.00	" 1.50	" 1.00
Girls " "	" " 2.00	" 1.50	" 1.00
Young Ladies Handicap	" " 3.00	" 2.00	" 1.00
Married Ladies Race	" " 3.00	" 2.00	" 1.00
Mens Free-for-all	" " 5.00	" 3.00	" 2.00
Ladies Coal Race	" " 3.00	" 2.00	" 1.00
Ladies Baloon Race	" " 3.00	" 2.00	" 1.00
Mens Coal Race	" " 3.00	" 2.00	" 1.00
Mens Sack Race	" " 3.00	" 2.00	" 1.00

Ladies' Tug-o-war-Gebo Ladies vs. Hot Springs County-Winning Team 22.00

Mens' Tug-o-war Gebo men vs. Hot Springs County Winning Team 22.00

Ladies Husband calling contest 1st. 3.00

Mens hog calling contest 1st 3.00

Ladies Rolling-pin throwing contest " 3.00 2.00 1.00

Horseshoe pitching contest " 12.00

Fat mans race " 3.00 2.00 1.00

Boys Smack-a-feller game Last two boys up 3.00 2.00

Lifting contest 3.00 2.00

Boy Scouts Game 5.00

Centipede Race 5.00

Gebo Wyoming

Opposite: The huge annual Labor Day celebrations included a variety of activities. Count Carey was on the boxing card as "Bobcat" Carey. *Mileva Maravic_6309_Box1_Folder Gebo Labor Day Broadside_1923, American Heritage Center, University of Wyoming.*

Left: The Labor Day activities were not just fun; they came with cash prizes as well. *Rephotographed by Levi Shinkle.*

racing. They also entered the rolling pin throw to see who could fling it the farthest. Because women did not wear pants in public at that time (or seldom owned pants), all of these activities were performed while wearing a dress or skirt and blouse.[217]

Marilyn Jones Revelle (Annie Kowlok Jones's daughter) related a story she heard about one Labor Day celebration. At one time, the big event during the celebration was the "fat ladies' race." The women raced from the Union Hall to past Burnell's house on the west side of town. The contestants wore simple house dresses. The men "placed bets all day long before the race started."[218]

When the event finally started, one lady took off and was well ahead of the others. Suddenly, she tripped over a rock and slid on her face and chest over the gravel road. She was carried home and laid out on her kitchen table. Her competitors brought rubbing alcohol and tweezers. It took several of them many hours to "pick little rocks out of the injured woman's skin."[219]

Two monstrous tug-of-wars were always held. The Gebo ladies competed against a team made up of Hot Springs County women. Likewise, the Gebo men went up against a Hot Springs County team. Festivities went on long into the night. Sometimes, after imbibing in the moonshine that was freely available, fistfights might break out. However, this was an accepted happening, and "the best man won."

What Do We Do Now?

When the Great Depression struck the United States in October 1929, the demand for coal began to diminish. Railroads had already begun to convert to diesel fuel. According to Rufus J. Ireland Jr., by 1932, the amount of tonnage produced had decreased to an unprofitable number.[220]

Owl Creek Coal Company closed for good in January 1938. The *Thermopolis Independent Record* reported:

> *OWL CREEK COAL COMPANY GOING OUT OF BUSINESS; WILL LIQUIDATE AND SELL OFF ASSETS*
> *Five Hundred Men Thrown Out of Work: Must Hunt Other Occupations or Locations. Mine Definitely Will Not Be Reopened. Coal Supply Said Not to Justify a Large Mine. Numerous Truck Mines May Supply Local Demand.*
>
> *In a statement to the* Independent Record *yesterday morning, R.J. Warinner, Gen. Manager of the Owl Creek Coal company, stated that the company closed its operations at the end of January, at which time it was employing about 500 men. There is no possibility, Mr. Warinner said, of the mine ever opening because its coal supply is exhausted.*
>
> *The company began operations at Gebo August 27, 1907. It has produced between eight and ten million tons of coal. The peak was 3,500 tons daily in 1929. The company was organized by Alfred Sully of New York, who took over the holdings of Messrs. Cottle, Gebo and others. When Mr. Sully died, ownership of the company passed to his daughter, Mrs. R.J. Ireland. Mr. Ireland operated the mine until his death, since which*

time his son, Rufus J. Ireland, Jr., has been its President. F.H. Burnell has been superintendent since March 1926 and Mr. Warinner, general manager since August 1926.

The Owl Creek Coal Company Mines closed permanently Monday, Jan. 31. That morning a notice was posted at the mine notifying the men that there would be no work as the mines were quitting business, definitely and permanently. To The Independent Record *Supt. P.H. Burnell stated that the company had run out of coal and had to stop. He said the company had prospected and drilled in many places over the county and had failed to find coal in sufficient bodies to justify installing another big railroad mine. He said there is a lot of coal but so broken up and located in pockets that it would not justify the Owl Creek Company to open other mines. He said the machinery would be removed from the mine, the store closed out, the buildings dismantled and the material disposed of. The company he said is definitely closing up its business, salvaging material and going out of business. Mr. Burnell said the bank is a separate corporation and will be operated just the same as before.*

Teachers at Gebo have not been paid since November, but were assured that their pay would be forthcoming and to continue the schools until May. The teachers have not been paid because the Coal Company has not paid its taxes, amounting to around $12,000. However the Coal Company gave the county treasurer assurances that the taxes would be paid.[221]

According to Mileva Maravic and Cyrene Fry Flohr, Dave King leased Mine #1 and continued to run it as a truck mine until 1944. It was renamed the Owl King Mine. Pat Burnell leased Mine #2 until the early 1950s and also owned four truck mines.[222]

Many of the miners moved their families away from Gebo immediately. This caused the high school population to decrease significantly. There were no longer enough students to justify keeping the high school open. Starting in the fall of 1938, the high school–aged students still in Gebo went into Thermopolis for high school. The high school portion of the school was torn down.[223]

Most of the miners who left simply abandoned their houses. Some of the residences were torn down. Others were moved to another location.[224]

Young boys would gather whenever a house was going to be moved to another town. It was interesting how they got the houses on to a truck with long large diameter logs connected to wheels in the back. When they

The King family leased Mine #1 and operated it as a truck mine until 1944. *Front, left to right*: Bud King, John King. Back row, *left to right*: Joe King, Dave King, "Hungary" Maxwell (?), Tony (Eli) Donozetovich (?).

> *were empty, you could see the truck attached to the logs which extended back far enough to support a house.*[225]

The houses that were left were sold for fifty dollars. Many miners took advantage of that opportunity. Mary Hanssen and her mother bought several of them to rent out. No one was allowed to resell them.[226]

Much of the infrastructure and many of the buildings around the mine were torn down. Since they were made of brick and set on concrete foundations, they could not be moved. Word went out that young boys could earn one penny for each brick they cleaned by removing the mortar. Each boy had to stack his cleaned bricks neatly in back of him so he could get credit for his work. At the end of the week, each boy was paid. The first order of business was to run to the pool hall for a candy bar, a double-dip ice cream cone or a bottle of pop. Since most of the boys' parents were not highly educated, the remainder was often deposited into a savings account at the Thermopolis bank for a college fund.[227]

Those miners who stayed were left to find new opportunities to make a living. Besides those who were able to work for King and Burnell, others worked in agriculture, the oilfields, construction or additional jobs. In every case, the conscious choice was made to stay and "make it work" in Gebo.[228]

The New Truck Mines

There was still a viable amount of coal to be mined in Gebo after Owl Creek Coal Company closed. Several miners decided on their own, or with others, to try to continue mining. These operations were called "truck mines" because a customer had to bring his own truck to the mine. Also, the seller could deliver it or the coal could be trucked to Kirby to be shipped out. Some of the mines included Roncco; Osborn; Coleman (which was bought by Rudy Kowlok and renamed the T-K); Haverlock and McCallum; Miller/Sheridan (owned by Peter Kewit & Sons; their trucks were bright red and were nicknamed the Red Devils); #4 and #5 (they had to cut down the size of the coal cars in #5 because the vein was so small); Ten X; and Big 6 (the latter four owned by Pat Burnell). Big 6 was so named because it had at least six feet of coal in its vein. Bill Deromedi's father and Louis Toth set a record of loading twenty-eight cars of coal per day in 1944–45 in the Big 6.[229]

One partnership formed in 1941 was that of John Henry Trusheim, Leo "Shorty" Roncco Sr. and Bart Cavalli. They formed the Roncco Coal Co., which was located west of Gebo.[230]

They were offered two options: purchase Mine #1 from Rufus Ireland or "buy 160 acres of OCCC land, lease C.W. Axtell's adjacent eighty acres, and buy the old White Mine property." The partners "did some exploring" and decided that purchasing the 160 acres and the additional land would be the best decision.[231]

The tipple operation of the Osborn Mine sorted the coal and readied it to be loaded into the trucks. *Jerry Deromedi.*

Another truck mine that started after OCCC closed was the Roncco Coal Mine. The partners built the Roncco Mine themselves. *Nena Roncco James.*

Left: Leo and Mary Roncco met in Gebo. She worked in the post office, and he checked his mail all the time. *Nena Roncco James, rephotographed by Levi Shinkle.*

Right: Henry and Susie Trusheim met in Gebo. She worked in the post office, and he "kept his name in front of her." *Janet Zupan Philp.*

Henry and Leo were brothers-in-law. Leo owned a farm south of Thermopolis in addition to his mining obligations.[232] He and his wife also sold food to make extra money. Her popular offerings included pies, buns, hamburgers, hot tamales and doughnuts. They did especially well when there was a dance at the pool hall.[233] Henry worked the land on Leo's property. He also ran a taxi service, because so few people had cars.[234] Bart worked construction after Mine #1 closed and helped build the new courthouse in Thermopolis.[235]

They started their company on July 4, 1941, with "a pick, a shovel, and a wheelbarrow" and built everything from scratch. Henry built and ran the tipple (which burned down and had to be rebuilt), Bart ran the underground operation and Shorty hauled and sold the coal.[236] Shorty's oldest granddaughter, Nena Roncco James, remembers going out to the mine with her father to see the mules, before the hoist was installed.[237]

According to Henry's granddaughter Janet Zupan Philp, he needed a new motor for the rebuilt tipple, so he and his son John "Jake" went to Denver to buy an old truck that had a motor they could use. They decided to drive the truck back to Gebo. However, that truck ran only in reverse. So Henry drove it all the way from Denver to Gebo—backward. Jake followed him with "red flags on sticks, hanging from the truck windows."[238]

The Henry Trusheim family is pictured in 1943. *Back row, left to right*: Carol, Josephine, Betty. *Seated, left to right*: Henry, Susie, J.H. Jr. *Janet Zupan Philp.*

When Henry's youngest child, Carol, was twelve years old, he needed to make a delivery. He wanted her to make the delivery with him to take coal to eastern Nebraska. It was a two-night run. Even though she was so young, she already had a driver's license. Henry drove the lead truck, and Carol followed him, driving a semi tractor-trailer.[239] Because she was so young, she could not see over the steering wheel. Henry laid boards across to raise the seat, which his wife covered with pillows to make sitting more comfortable. Henry attached blocks of wood to the pedals so Carol's feet reached them and she could shift to another gear.[240]

Carol told her father she didn't know when to shift. He explained that she just needed to "watch his exhaust." When she saw the exhaust change color and make a "puff," she was to shift. He also told her to listen to his truck to hear the shifting. However, the trucks were so loud that it was doubtful she could hear anything. Lastly, she was told to "watch your side of the road, and don't get in the snow." When they stopped for any reason, she was not to get out of the truck without her father. She even had to sleep in the truck both nights.[241]

The John Kowlok family came to America from Poland and eventually settled in Gebo to work in the mines. Their son Rudy worked in the Crosby mines and then in OCCC Mine #1 until it closed. He worked for the Burnell truck mine and went to Rock Springs, Wyoming, during World War II. He came back to the mines in Gebo in 1944. He worked for the Sheridan Coal Company from 1944 to 1952 and then at the Roncco Coal Company from 1952 to 1957. In 1957, he took over the Coleman Mine. George Talovich was a partner for a brief time, and then Rudy opened the T-K mine. At the same time, he purchased the truck mine tipple, which was in Kirby.[242]

The mine was located west of Gebo, up Little Sand Draw. His son H. Rudy worked in the mine as well. At first, Rudy used a chain conveyor to bring the coal out of the ground. Later, the coal was brought out of the mine in cars on tracks. He bought a hoist in Rock Springs, Wyoming. It was several tons in weight, and he went to get it in a two-ton truck, but he still managed to get it back to Gebo.

The slope of the mine was from seven hundred to nine hundred feet deep, and the miners walked down the nine-hundred-foot main slope. They entered rooms that broke off on both sides of the slope. In order to expand the mine, the men started on the bottom (called the solid) when they were driving slope (going down), and they shoveled right into the waiting coal cars. This was the most difficult work. Most of the coal came from the rooms. It was shoveled into the pan line (which looked like a rain gutter, only bigger) and shaken down to the cars, when they were in the rooms. Cardox shells (a carbon dioxide blasting system) were used to dislodge coal into chunks. The holes for the shells were two to three inches in diameter and three feet long, which gave bigger lumps. Rudy liked that. The air shaft was connected to the fan to provide ventilation.

Rainstorms happened often, and the narrowness of the draw caused flash floods to damage the road (which had to be rebuilt often) and flood the mine. The oilfields west of the mine became a source of the culverts used in the mine road: old oil tankers. One day, after the family had moved to Kirby, a severe storm came up. Rudy and H. Rudy tried to drive home to Kirby, but the road was washed out. They ended up walking nine miles back to Kirby and got home very late after dark.

Crew sizes depended on time of year and demand at that time. There were usually four to five miners "down in the hole" and one miner up on top running the hoist. They normally filled three to four ten-ton truckloads per day. Those underground made an hourly wage, but the hoist operator may have made less. They worked five days a week, and the mine was open until 1970. "They were hard workers and had integrity," H. Rudy said.[243]

Little Sand Draw is located near the location of the T-K Mine. *Photograph by Levi Shinkle.*

The T-K hoist is still visible today. *Photograph by Levi Shinkle.*

The family lived in Gebo until 1963, when they moved to Kirby. The tipple had burned down, and they rebuilt it. Rudy's wife, Evelyn, ran the tipple office and later the Kirby post office. There were seven Kowlok children: Ken, Jerry, Miriam, Dolores, Carol, Judy and H. Rudy. They lived in a small house near the Gebo school. With that many people in the house, plus her duties at the tipple office, Evelyn was always busy. The one chore that took a huge amount of time every day was laundry. It was a true "hardship."

The youngest daughter, Judy, had a set of cooking pans with which she loved to play. One day, she and her friends decided it would be a good idea to build a fire outside so Judy could "cook." Unfortunately, the fire was built under the clean laundry that was drying on the clothesline. That load of laundry had to be rewashed, and Evelyn was *not* happy.

Ken and Jerry joined their cousins Gary, Phil and John Giannino and took turns going with Grandma and Grandpa Kowlok to the Big Horn Mountains after the coal seasons. They ate fish three times a day, plus chokecherries (jelly and wine) and game meat. There were pancakes and jelly to take along when going to fish. The rule was that the previous catch had to be completely eaten before anyone in the group could go fishing again. Gary disliked fish, so he kept trying to give away the fish he had caught to tourists who were in the mountains. "Want some fish?" he asked hopefully, over and over again.

H. Rudy was born in 1949, went to school in Gebo for the first grade and then went to Lucerne when Gebo school was closed and consolidated with Lucerne. There were only four grades in Gebo that last year (1955). At home, he and his friends loved being outside:[244]

> *The kids played around the rocks. When it got too hot in the afternoon to be out (because the snakes hid under the rocks to stay cool), everybody came in and played board games. My friends in Gebo were Deromedis, Urists and Hendryxes. When I was older, my friends in Kirby and I played "Rat Patrol" in an old Jeep; Kick the Can; Hide and Seek; Red Rover; and baseball.*[245]

Barbara "Bobby" Heron Workman was born in Thermopolis in 1938. Her family moved to Gebo when she was in first grade, around 1943. Her father worked for Mine #2, which was west of Gebo and run by Pat Burnell. Her dad drove the mine's truck down to Kirby so the coal could be shipped out. Occasionally, she and her younger brother Larry walked to and from school on the highway that went to the mine. They crossed the tracks when they got to town.[246]

There were just two classrooms in use in Gebo School when the Heron children attended, as the high school had been torn down. When they got to third grade, the students were old enough to walk up to the Gebo Bar at lunch to buy candy bars. They had to go to the back sliding window, so they wouldn't enter the bar itself. So many students made that short trip every day that a path was worn between the two buildings.[247]

Their house by the mine was "nice" and had a wood floor. One time, when their parents both got home at 6:00 p.m., Bobbi made a snack for herself and Larry. She thought she could make toasted cheese sandwiches on the big cast-iron stove. Her mother left coals in the fire box, so Bobbi put in kindling and got the grate hot enough. She put Velveeta cheese between the bread slices and grilled them carefully. That night, her mother was very concerned, because she had cautioned both children about starting a fire. Bobby assured her that she had not started a fire, just "put enough kindling in to cook the sandwiches."[248]

The family lived in Crosby one summer when Mr. Heron was between jobs. The fires from the Crosby mine explosion were still burning. The Heron "house" resembled a couple of railroad boxcars put together. It had four rooms and a lean-to on the back of the kitchen for storage. It was not a standard house. The camp Crosby provided the house for them.[249]

Left to right: Kenard, Eddie, Larry and Barbara "Bobby" Heron are shown outside their home in Gebo. They were getting ready to go to the end-of-the-year school picnic. *Bobby Heron Workman.*

The family that lived in the house before the Herons must have had a young girl, because a "little doll bed" had been left. It was fashioned out of a cookie sheet for the mattress, spools for the bed frame and baby blankets to keep the dolly warm. Bobby loved it and played with it all the time—until her mother needed the cookie sheet for baking.

In Gebo, they also lived in their friend Pete Solich's house near Deromedis, Cavallis and Randalls. There was a garage that had been dug out of a hillside nearby. It wasn't used as a garage at the time, but the kids loved to play in it, especially because it was so cool in the hot summer.[250]

One of Bobby's favorite Gebo memories came from the time several of the neighborhood families got to go to Thermopolis to the Disney musical movie *Song of the South*. It starred James Baskett (Uncle Remus), Bobby Driscoll (John) and Hattie McDaniel (Aunt Tempy). Each family went separately, but everyone was so pleased when they saw the others there. Going to the movies was a real treat for all. The show was in Technicolor, which Bobby remembered as "amazing." The songs that were performed delighted them all, and when they got back to Gebo, the children all got together and sang their favorite songs over and over.[251]

Bobby absolutely "loved" living in Gebo.[252]

Later School Happenings

Duke Bolich shared another powerful memory from school:

> *It was springtime in third grade. We all looked forward to going outside for recess when it was a nice day. The girls would jump rope on the grounds to the East of the front of the school. The boys would bring their marbles from home and spend the time "playing marbles."*
>
> *In the first few minutes of the time we were outside, I lost two games and had no more marbles with which to play. I had kept some others in my desk in the classroom so I ran back into the room, opened up my desk, and grabbed five or six marbles. As soon as I put the marbles into my pocket, I ran for the door to the room and dashed out to the outside foyer door. I did not bother to close either one, as I was focusing on getting back into a marble game and winning some of my marbles back.*
>
> *We were gathered around a ring that was marked in the dirt with an index finger of one of the players. Each player would throw his marbles inside the ring. We concentrated on being able to take a marble in our hand, hold it in our index and middle finger and shoot it out with our thumb. We became quite good after a lot of practice. If you hit a marble inside the ring and knocked it out of the ring, you got to keep the marble. If you knocked a marble out of the ring you got to have another shot. If you did not hit a marble and move it out of the ring it was the next boys* [sic] *turn.*
>
> *Just when we had a great game going, Miss Hanssen became very alarmed, because a sheep herder was driving about 500 sheep through the coal camp and then passing by the front of the school. Since I had not bothered to close either door, the lead sheep headed toward the foyer door and went into the school. As animals often do, the following sheep moved into the foyer and the classroom.*
>
> *Miss Hanssen started yelling at the sheep to get out, but that just spooked them even more. Several sheep made it into the classroom. We all came running to get them out of the room and back with the rest of the pack. This*

made the sheep even more frightened, because we were all yelling for them to get out. It must have been some kind of sight, with the teacher screaming, 20 or so kids chasing the sheep, and the sheep frantically looking for a way out of the situation.

The panicked sheep pushed all of the desks out of their normal position in the classroom, and knocked whatever was on top of the desk onto the floor. With the help of everyone, we finally got the sheep out of the room and outside to the rest of the herd.

The sheepherder apologized to Miss Hanssen and moved them past the school. As soon as everything became normal again, Miss Hanssen asked, "Who left the doors open?" I raised my hand in fear. She came over with her ruler and gave me about ten swats across my open hand.

I was really embarrassed because I had never gotten into trouble in school and always tried to be a good student. To say the least, I really got teased by the other kids in my class and the 4th graders. The first and second graders didn't say much but laughed about the mess we had to clean up. Miss Hanssen never referred to my "spanking," but I made sure I always closed the doors behind me from that day forward.[253]

Later Camp "Services"

The Urist family owned the water company in Kirby for many years. Twins Patsy and Patty were expected to do their part to help the business:

Patsy (Urist) Dorman (*left*) and Patty (Urist) Lebolo (*right*) shared many childhood adventures in Gebo. *Patsy Urist Dorman, rephotographed by Levi Shinkle.*

The Urist family were all involved with the Water Works. When Patty and I were nine or ten, we helped as well. We had to dig for water leaks. When the water lines were dug from Kirby to Gebo, cow manure was packed all over and around the pipes. The manure kept the lines from freezing. Then dirt was packed around the manure. Over the years, the manure ate holes in the pipes. If the water needed to be turned off, it took time for the water tank to fill back up. When the water went back on, the pressure had built up and blew holes in the pipes. In order to find the leaks, Patty and I looked to see where Sig Nelson's cows were congregated, which meant there was probably a leak. If the ground was frozen, we had to set tire fires to thaw the ground. In every case, we dug with pick, shovel, and bucket. We had to make a hole six feet long, three feet wide, and six to eight feet deep. This was done in every season of the year. Our dad put a wooden plug in the hole and then wrapped a layer of inner tube around the pipe. A welder had to come and put the replacement pipe on. We hoped it would hold.[254]

ADDITIONAL ACTIVITIES AND OPPORTUNITIES

Boy Scouts

In those days, each Boy Scout chose a specialty skill he was most interested in. Count Carey really liked semaphore signaling. This entailed using two red and white flags on staffs to make patterns that stood for letters of the alphabet. Count's partner, or receiver, was Si Ashley. The receiver was responsible for "reading" the signals, or words being sent, and reacting accordingly. Count and Si competed against other troops around the state and won several trophies for sending and reading the signals the fastest and most correctly.

Spending time outdoors was, and is, a very large part of the Boy Scout experience. Count learned his appreciation of camping, fishing, hunting and surviving in the outdoors during miners' strikes. During the strike of 1920, the Carey family spent approximately three months on the Wood River near Meeteetse, Wyoming. The majority of families in Gebo took similar trips during strikes. Count learned many outdoor skills during these family outings.

Mr. Jones also taught the boys first aid through the Mines First Aid Course.[255]

Civilian Conservation Corps

After the OCCC mines closed in January 1938, many Gebo families were left with few options to keep making a living: move from Gebo and get work in another mine or stay and try to find a reliable source of income. In March 1938, nine boys from Gebo became "additional breadwinners" for their families when they joined the Civilian Conservation Corps (CCC), assigned to Gillette, Wyoming. The group included Robert Pisto, William "Billy"

Maxwell, William "Bill" Radulovich, Mike Novakovich, Ted Novakovich, Richard "Dick" Lathum, Joe Cavalli, Julius Turner and George Talovich. George (who was Mileva's stepbrother) shared the following story.[256]

After the Great Depression began in 1929, the U.S. government instituted the CCC. Its mission was to provide a means for young men nineteen years and older to earn money and to preserve natural resources in the areas where they were assigned. The Gebo group was tasked with fighting coal mine fires in Gillette out of Camp GLO-1-W. They also fought forest fires around Devil's Tower.

They rode the train from Kirby to Billings. While there, Julius Turner became homesick. He "went over the hill" and returned to Gebo. From Billings, the rest of the group took another train to Gillette. Even though it was a nice spring day, there was a foot of snow on the ground.[257]

When they arrived at Camp GLO-1-W, they were first housed in a barrack called the BARN. That night, they had to experience "being initiated" at 2:00 a.m. Their beds were turned over (with the boys still sleeping in them). Because it was the middle of the night, the barrack's coal-burning stove had gone cold. It was very jarring to be thrown out of a warm bed. The next morning, they were assigned to a permanent barrack and issued their CCC clothing.

Their camp had one main mission: extinguish existing coal mine fires. The mines they were assigned to had been named the Ditto, the Laur and the Sparta coal fires. The first two required them to terrace and turn over the ground to allow them to "shut off the oxygen that was fueling the coal fires."[258]

The Laur project necessitated the building of walls in the mine to "shut off the oxygen" and "removal of the fire with a big shovel which they loaded into trucks." The hot ashes and burning coal caused the coal trucks the boys were loading to become extremely hot. The trucks had to be "shut down several times during the day."

At the Laur project, because the boys were working in the mine itself, they could only work half a day at a time. The black damp and bad air caused everyone to become ill.[259]

The boys earned one dollar a day, thirty dollars per month, for their efforts, plus meals, housing and clothing. The government sent twenty-five dollars to the families, and the boys got to keep five dollars for their "personal expenses." They paid twenty-five cents to go to the movie and one-eighth of a cent to play a card game called Chinese Fan Tan. It was their top entertainment.[260]

Because the boys joined the CCC in March, they could not finish their school year. An agreement was made between Gebo High School and the government that the boys' coursework would be satisfactorily completed. Their assignments were sent to Mr. Carlson, the boys' counselor at GLO-1-W. He made sure that the boys attended classes, completed all their studies and passed all their tests. Every one of the boys completed his studies and passed the school year.

There were free-time activities. The camp had a "great" softball team. A Nebraska boy named Hart was an amazing pitcher. The Gebo boys were also grouped with Texas and Oklahoma boys. All the boys helped take care of the camp's mascot, a black bear named Mae West.[261]

It was "the hardest work they ever did," but they enjoyed it and were very grateful for the chance to earn money. Their foremen at the projects were all ex-miners. They treated the boys "like they were adult miners when they were just teenagers." The Gebo group "saved many coal beds in the Gillette area from burning by extinguishing the fires."

Because they all completed their classes for Gebo High School, they returned to Gebo in September for the next school year. As the high school closed at the end of the 1938 school year, all of the Gebo High School students went to Thermopolis High School for the remainder of their high school careers.

When the boys were seniors, several of those who were in the CCC played football for the Thermopolis High School Bobcats.[262] A second group from

Three of the boys who joined the CCC in Gillette played football for the Thermopolis Bobcats in 1939. *Front row, left to right*: Hank Cabre, Bill Radulovich (CCC), Bill Bush, George Talovich (CCC) and Joe Cavalli (CCC). *Lea Cavalli Schoenewald.*

Gebo joined the CCC in 1938 and were sent to Colorado, including Eugene Carey, Eddie Todorovich, Mike Maravich and Eddie Cunningham.[263]

In 1933, the CCC sent Pete Diemer, "Pop" Maxwell and Val Cammock to Encampment, Wyoming, and then to Denison, Texas. The next year, Mike Maravich, Eddie Todorovich and Anton Bobeck worked in Jackson Hole, Wyoming; and Jack Woolman, Jack Wilson and John Kowlok went to Lander, Wyoming, first and then to Jackson. In 1937, Bob Johnstone was part of a group that fought the Blackwater Fire between Cody, Wyoming, and Yellowstone National Park.[264] The fire spread rapidly and created a firestorm. Fifteen firefighters from another crew were killed, including several CCC boys. Bob's crew had to go in and recover the bodies.[265]

NO CHOICE BUT TO CLOSE

The demand for coal continued to drop. Eventually, the truck mines began closing. Henry Trushein was bought out, and Leo Roncco Jr. and Pete Cavalli took over the Roncco Coal Company from their fathers. They

Leo Roncco Jr. follows a full coal car up to the tipple at the Roncco Coal Company. *Nena Roncco James.*

The Mine #1 fan stands as a reminder of the lost district that once flourished. *Photograph by Levi Shinkle.*

were the last to stay in business. In 1975, the Mine Safety and Health Administration (MSHA) imposed several new requirements on small mines. The cost was prohibitive, so this next generation also was forced to close. Dorothy Buchanan Milek estimated that there were over forty mines that had operated in the county.[266]

In 1971, the Bureau of Land Management (BLM), which had taken over the entire area where Gebo was located, deemed it "dangerous." There were too many opportunities for injury, with all of the abandoned buildings, remnants of the mine operations and other debris. It was decided to level the area. Former Gebo residents, especially, were extremely upset with this decision. Additional reclamation work was done in the early 2000s to cover erosion from cave-ins and refill entrances and holes.

According to Ray Shaffer, Gebo was the largest town in the Big Horn Basin at its peak (even though it was never incorporated). Today, all that remain are a fan from Mine #1; some scattered foundations; partial walls from the Rock Row homes; the cemetery; various dirt roads; and a few interpretive signs placed by the Bureau of Land Management. The signs contain pictures of the thriving camp, with important landmarks described.

The area is a popular destination for target practicing, artifact collecting, hiking, four-wheeling, ghost hunting and other activities.

> *Low foothills, an expanse of prairie, miles in the distance, the eye beholds the hazy blue line of mountains, this frames a typical, small picturesque Wyoming Coal Town, in description—my hometown.*
> *—Mileva Maravic*[267]

CHAPTER 4

KIRBY

Founding and First Businesses

Kirby's importance in and to the district cannot be overstated. The railroad made it possible to ship out the amazing coal found in Crosby and Gebo. The Big Horn River provided the water needed by everyone and everything. And the Kirbyites became the "survivors"—those brave souls willing to "stay the course," even when the mines closed and uncertainty ruled their futures.

In 1892, John Nelson desired to ranch on the Big Horn River, several miles north of Old Town Thermopolis. He filed for a homestead. In 1907, his family's plans changed, as the railroad entered their world. The Burlington Line, which became the Chicago, Burlington and Quincy Railroad (CB&Q), purchased acreage from Nelson, and his homestead evolved into a townsite named Kirby.[268]

However, the settling of Kirby did not go smoothly at first. In September 1907, the *Wyoming Tribune* reported that there were no town lots being sold. No buildings were going up, with the exception of a depot and warehouse on the railroad right-of-way. The townsite was "in litigation," but permits to build were expected to be available shortly.[269]

Eventually, Nelson built a bar and café and a twenty-room hotel. The town was named for Kris Kirby, a cowboy who "ran cattle in the area." The first home in Kirby was built by the Andrew Johnson family.[270] Freight bound for Thermopolis was shipped into Kirby and taken in by horse and wagon. Furniture for the new Emery Hotel was handled by the Mayfield Freight Team.[271]

Left: The railroad infrastructure expanded rapidly.

Below: The Kirby Depot was an important part of the railroad's presence.

By 1908, the town was booming. A second family home, built by J.B. Wight, also included a grocery store and post office run by him and his family (off and on) until the 1920s. Other businesses and amenities that were offered included A. Johnson's mercantile, a saloon, a stable and stage line, J.A. Swenson's store, a public hall and a baseball field. A business "district" had taken shape.[272]

In 1909, almost all of the businesses went up in flames when Swenson's new store burned. But all of them rebuilt and were joined by many others in the coming years, including a Greek bakery; John Vinich's bakery, run by Joe Obradovich; Bob Smiley's pool hall; Billy Hergert's garage (which replaced the livery stable and stage lines); and F.W. Bull's carpentry shop. In addition, the Nelson brothers, Ed and John, and their partner Phil Horr bought Swenson's store and enlarged it; Sig Nelson, John's nephew, worked

Top: The Kirby Hotel welcomed guests for many years.

Bottom: The business district of Kirby grew quickly once the first buildings were finished.

for his uncle and then took over the business, delivering groceries to Gebo and Crosby; Ferko and Tatarko took over from A. Johnson at the mercantile; and numerous bars were owned by A. Johnson, Bergloff and Talovich, Gus Sylvester Taylor and Dan Skoric.[273]

It became clear that a fully functioning city government was needed. On August 16, 1915, the town was incorporated by a vote of sixty-one to one. A. Johnson was voted as mayor, and the city council elected included John

Sig Nelson made grocery deliveries for his family's Kirby store.

Nelson (who also served as treasurer), J.B. Wight, Carl Braemer and John A. Stuart. The new mayor appointed Phillip Horr as clerk and police clerk and E.H. Jenkins as marshal.[274] Numerous emergency ordinances were passed, approved and filed in order to begin the process of governance:

> *Corporate Seal; Order of Business, Public Peace and Safety; Concerning Misdemeanors and Providing Penalties Therefor* [sic]*; and Concerning Appointive Officers, Prescribing their Terms of Office and Fixing Their Salaries; Concerning the Providing for the Granting of Licenses to Engage in and Conduct the Business of Retialing* [sic] *Malt and Vinous and Spirituous Liquors; Concerning the Creating of the Office of Police Justice and Fixing His Duties and The Term of His Office.*[275]

A jail was needed and planned, and the contract was recorded on October 4, 1915. The building construction cost was $275. It also served as the town hall.[276]

The Kirby jail was the first, and only, building constructed by the town. *Lea Cavalli Schoenewald.*

Paul Klos and Lefty Graham, who both grew up in Kirby, were interviewed about their hometown. They recalled one of the most well-known "businesses" in Kirby. Eventually, east of the jail/town hall was a "most notorious" structure, a "house of ill-repute" owned by Gertie Harris. "This was a rough little town once," said Klos. Graham agreed and added, "Crosby and Gebo were company towns, but the men who lived there didn't have establishments to entertain themselves as Kirby did. Consequently, they visited Kirby frequently, which at the time had about nine open saloons and gambling joints besides the house of prostitution." In 1919, the mayor of Kirby suggested that "a fine of $100 'be assessed on the house of Gertie Harris to be paid by her each month…to cover all fines in connection with the running of that house, number of inmates not to be considered.'"[277]

In the first days of Kirby, the marshal earned from $25 to $50 a month. When Klos served as marshal, he made $200 per month. He had to jail people only three or four times, and only because of fighting and drunkenness. He started on a "call basis" but then patrolled the streets looking for "youngsters shooting b-b guns or throwing rocks."[278]

Government Over the Years

The Kirby Town Council faced and weathered a number of situations and crises during the days it was part of the district, as evidenced by its meeting minutes. After its incorporation, numerous ordinances were needed to function as a governing body. These included voting to grant Owl Creek Power Company a franchise to provide water to Kirby. They also granted the utility a right-of-way for water mains and pipelines. Later, in 1919, there was a contentious vote to spend $130 on "Base Ball suits." The mayor broke the tie and voted in favor.

In the 1920s, many of the meetings were devoted to discussing and paying bills. Taxes and fees were also a priority. Owners of female dogs had to pay $5.00, and male dog owners were charged $2.50. The marshal was instructed to put up notices warning against speeding and also warning children not to "catch rides on autos in the streets." At the end of the decade, it was necessary to prohibit "any excavation, dam, waterway, ditch reservoir, and depression upon any street or alley." Any violation would result in "punishment by special assessment."

The 1930s brought more discussions about various bills and selling of licenses. Liquor license applications were often considered, as were applications to sell "3.2 beer and other beverages containing less than 4% alcohol." During this decade, the council voted to appoint women as clerk and treasurer. After the Gebo mines closed at the beginning of 1938, there was no mention of that situation during all of that year. In 1939, the topic of water occupied many meetings. The mayor was asked to contact Mountain States Power about providing both water and power to the town. Next, a "Mr. Wariner [*sic*]" offered "an entire pumping plant, all equipment, etc. to the town for $1,000. A committee was appointed to investigate." They even discussed buying the Gebo Pipeline.

In the 1940s, more taxes were discussed and voted on. Also, World War II brought the desire to help the war effort. The tax levy for 1940 was $500, and for 1943 it was $300. Liquor licenses cost $375. The town bought a Defense Bond in February 1942. Three new council committees were instituted: Department of Public Affairs and Safety; Department of Finance, Accounts and Public Property; and Department of Streets and Public Improvement. In 1947, it was reported that there were "facilities in town for new industry." On March 1, 1948, Sheridan Coal and the Haverlock and MacCallum Mine donated a coal truck and several men, and many Kirbyites "donated their services on Main Street because the street was badly washed out. They hauled 10 loads of gravel."

The early 1950s brought the need to buy insurance for the council. On May 7, 1951, the treasurer was instructed to "buy a blanket insurance policy to cover any accidents of any sort that the town officials feel necessary to cover."[279]

THE RAILROAD

As Kirby was the end-of-track for the line that eventually went south to Thermopolis, through the Wind River Canyon and on to Denver, Colorado, any railroad and/or local Kirby citizen news was an important part of the *Thermopolis Independent* newspaper. In December 1907, the column "Kirby Kernals" carried the following news items:

> *Work on the spur to the Stine & Crosby mine is making good progress; The section house at this place will soon be completed. Plasterers are at work this week; A special train of four coaches came up the line Thursday, bringing in officials of the railroad company; Several of the railroad employes* [sic] *went up to Thermopolis to see the Springs Sunday morning, returning to Kirby in the evening; The depot will be finished in a short time, which will be a great convenience to the people coming down on the stage, who have to wait while the train goes to Gebo and back; The Nelson hotel will be finished by the first of next month. It is reported that the train schedule will be changed to come through to the end of the line without any stop over night* [sic] *in Worland. The hotel will have ample room to accommodate the people coming in on the train.*[280]

In an oral history given to the Washakie County Museum in Worland, Wyoming, Ruth Kennedy remembered when the railroad came to Worland and then went on to Kirby:

> *When the railroad came into Worland, it was just like a "Y"—it was the end of the railroad and so they could back up. They just had small trains back then and they would go clear up to the end and then just back around and got it so that the engine could go back the other direction. And that was in 1906 and then in 1907 the railroad went up Kirby so they also built a "Y" up there.*[281]

The railroad eventually expanded its presence in Kirby with a roundhouse, a section house, a depot, various livestock-loading chutes and a wool warehouse. The people who kept the railroad going lived in "boxcar houses."[282] Working for the railroad exposed the men to dangers of many kinds. In 1923, one of the train engines exploded in Kirby. Two men were killed, and one man was seriously injured.[283]

At the same time that railroad traffic from Kirby south was being discussed, Asmus Boysen and his partners submitted a permit application to build a sixty-foot power dam approximately sixteen miles south of Thermopolis. The dam was to be located at the southern end of Wind River Canyon. Without an agreement between the railroad and the Boysen group regarding the location of the rail line, such a line might not be feasible.[284]

Additionally, the Burlington Line was not the only railroad with an interest in the canyon. The Chicago and Northwestern already had track built south of the canyon, and the company wanted to make sure it had access to any track going north through the canyon as well. Although headlines had been reporting that there was a "feud" between the two rail companies, they had actually reached an agreement regarding track use.[285]

Both companies met in Cheyenne with the state engineer and made it clear that they wanted to ensure the smooth flow of traffic through the canyon. The plan they endorsed included the Burlington Line building through the canyon. Upon reaching the south end, the line would "swing to the east," join the North Western (N-W) Line at a to-be-determined location and then continue on the N-W tracks to "Orrin [*sic*] Junction." The N-W Line agreed to build a short spur from a point on their existing track (out of Shoshoni, Wyoming) to the Burlington Line near the canyon. Thus, both lines could use the same track through the canyon and on north to Kirby. The N-W then had access to the significant coal fields. It was further reported that as soon as adequate financing was obtained, the "building will proceed."[286]

In the same edition, also on the front page, another article gave additional information regarding the importance of the railroad continuing south from Kirby.[287]

It was felt that every Wyoming citizen should be concerned about Asmus Boysen receiving his permit to construct the Boysen Dam. The dam and the canyon were located in the Owl Creek Mountain range, which branches from the "main range of the Rockies on the west." It then "swings southeasterly" to connect with the Big Horn range on the east. The elevation of the Owl Creeks gradually drops from a high of eleven thousand feet to six thousand feet. This completes the "chain of mountains that forms a natural barrier, cutting off all of Big Horn and a part of Fremont County from the rest of the state."

Exhaustive (and repeated) surveys proved that there was no viable route between the two counties except through the canyon, which is twelve miles long and rises to an average height of six hundred feet. The surveys also revealed that eleven tunnels would be necessary to pass through the canyon.

Residents of Big Horn County, in order to travel to Cheyenne, the capital of Wyoming, had to travel "north into Montana, then southeast through Montana, Wyoming, South Dakota and Nebraska, and thence west through Nebraska to reach their destination." This trip covered eight hundred to one thousand miles and took several days by stage. If they rode the train, it would take approximately ten hours. Even traveling to Shoshoni required a fifty-mile stage ride over a nearly impassable mountain. A railroad connection would be much more desirable.

Not only would the N-W have access to the highly prized coal near Kirby, but it would also be able to enter the various markets of Big Horn County. This would allow healthy competition and bring down freight rates. The highly profitable agriculture sector in Big Horn County would have access to additional markets. The railroad would also give the county residents another means of transportation. More and more portions of Wyoming would be linked.

The opportunity to carry to market the highest-quality coal near Kirby was another bonus for Big Horn County and the N-W. The coal was "equal to the best in the state, not excluding Rock Springs." There were coal veins all over the county, but the best deposits[288]

> *extend from Kirby on the east to Meeteetse on the northwest, a distance of about 60 miles. The principal vein is 14 feet in thickness at Kirby, and gradually increases in size as it extends westward, until it reaches the magnificent extent of 50 feet in thickness on the headwaters of Gooseberry creek. This coal averages about 5 per cent fixed carbon, 8 to 10 per cent moisture, 35 per cent volative oils and gases, 3½ per cent ash and ½ per cent sulpher* [sic]. *It stands storage, gives general satisfaction both as a domestic coal and for steam purposes, and, at a glance all will concede that its future market lies in the thickly settled Missouri and Mississippi valleys. At present there are two mines in operation at Kirby, one owned by the Kirby Coal company and the other by the Owl Creek Coal Company. The only market for the product at present is local in Big Horn county and to the northwest in Montana and other states, where it must compete with local Montana coals, having a much shorter haul and correspondingly lower freight rate. The coal industry in Big Horn county*

It was a "backwards" view for the engineer of this Kirby locomotive. He had backed all the way to Thermopolis, as he couldn't turn around there in order to return to Kirby.

> *will not reach its maximum until there is furnished a southern outlet by rail through the Wind river canyon, by which the product can reach the eastern markets by direct route.*[289]

The Burlington Line finally reached Thermopolis in 1910. Because Thermopolis became the new end-of-track, the train from Kirby had to back up all the way to the Thermopolis depot.[290]

In later years, railroad strikes became more prevalent. In 1922, the CB&Q advertised for needed workers in the *Thermopolis Independent* "to replace men on strike against decision of the United States Railroad Labor Board." The company marketed itself as "the West's most dependable railroad operating 9,389 miles of road in the 11 great wealth-producing states between the Great Lakes and the Rocky Mountains." It needed "well appearing men of experience" to work in Illinois, Wisconsin, Iowa, Missouri and Nebraska. The permanent positions and salaries available included "Machinists, boilermakers, and blacksmiths, 70 cents per hour; A few helpers for these crafts, 47 cents per hour; Passenger car carpenters and repairers, 70 cents per hour; Freight car carpenters and repairers, 63 cents per hour."[291]

The railroad was particularly interested in hiring seasonal farm and related workers for positions as helpers in the car and locomotive departments. Excellent efforts in those jobs would entitle the workers to qualify for higher-paying salaries.[292]

The CB&Q finally settled with the unions, and expansion was again possible for the line. Kirby's contributions to that expansion throughout the years allowed growth in many industries and areas of the entire country.

THE YELLOWSTONE HIGHWAY

Kirby was needed again, this time as part of the Yellowstone Highway. In 1915, a "well-marked, well-maintained highway" became a necessity in north central Wyoming.[293] That year, cars were allowed in Yellowstone National Park for the first time, and Rocky Mountain National Park was established northwest of Denver.[294] Eventually, the Yellowstone Highway would become the first section of the National Park-to-Park Highway. This roadway would connect all of the national parks in the western United States.[295]

In 1917, the State of Wyoming established a Wyoming Highway Department. As the department began studying the various routes throughout the state, it became clear that traveling from Casper to Thermopolis by car was a harrowing experience. Travel south of Thermopolis brought the driver face to face with numerous trails marked with yellow rocks that were supposed to mark the best route over the mountains. Frustration generally followed. The Highway Department decided that "a road must be built through the Wind River Canyon, on the opposite side of the river from the railroad route."[296] This section continued north out of Thermopolis, traveled parallel to the CB&Q tracks, veered right (east) at Coal Draw to just west of the railroad tracks, turned north to Kirby, ran through the town on what is now Bryan Street and continued for nine-tenths of a mile until merging with the present-day Highway 20.[297]

Later, the highway was realigned and bypassed Kirby.

A KIRBY MINER

John Bury was born in Istebna, Poland, in 1890. He came to America by boat (as a stowaway) in 1903, settling in Texas. His uncle managed to sneak him off the boat. He then went to Chicago to work in the stockyards. Cousins

Left: Anna Wagner Bury (*left*) and Sue Rushin Einer (*right*) are fashionable in Kirby. *Bury family.*

Below: The Bury children show off their finery in Kirby. *Left to right*: Betty, Mary, John Jr. and Henry. *Bury family*

in Wyoming eventually helped him with travel arrangements to go out west, and he settled in Sheridan, Wyoming.

John's future wife, Anna Wagner, was born in 1905 in Pennsylvania, one of six children. Her parents moved to Dietz, Wyoming (near Sheridan), where her father worked in the mines. He died, leaving her mother, Mary, to raise the children alone. In order to support her family, she took in boarders, did sewing and was a janitor for the school district.

Anna's mother remarried, to Mike Rushin. They had six children: Susie, Helen, George, John, Joe and Alex. They lived near Sheridan for several years. Mike farmed in the summer and worked in the mines in the winter. In 1925, they moved to Gebo, where Mike worked in the mines until an accident forced him to quit in 1934. They moved to Thermopolis, where they bought property that allowed them to raise vegetables to sell, as well as chickens and goat milk. Mike also worked on Works Progress Administration (WPA) projects for many years.

One of Anna's brothers, Andrew, died in one of the Gebo mines. He was a rope rider, the most dangerous job in the mine. The car he was riding separated from the cars behind it and went careening down a steep slope. It was thrown off the rail,[298] killing Andrew.

John and Anna married in 1918 in Sheridan; Anna was just thirteen. They moved to Kirby in 1928, and John worked in the Gebo mines. They had seven children: John Jr., Henry, Mary, Betty, Patricia, Francis and Dorothy "Dolly."

Because the mines became unprofitable in 1932, John's position was eventually eliminated. The family moved to Thermopolis in 1935, and John worked for the WPA for a time. He was on the crew that built the stone wall around the rodeo grounds, and he also worked on the stone walkways and structures by the Big Spring in Hot Springs State Park. The Hot Springs County Road and Bridge Department was his next job, and then he became a custodian for the school district so he could make more money.[299]

Schools

The school population was always representative of the coal mining and railroad industries. The first Kirby school photo was dated 1908. Two of the children who attended were Holly and Clarence Kelly.

The first school term in the incorporated town started in 1913. The classroom was a railroad boxcar taken off its wheels, located north of town,

The first school building in the town of Kirby housed these students in a boxcar taken off its wheels. The teachers who worked there included Bertha Ryan, W.J. Sisk and Ada Kobelin.

on the railroad Y. It lacked adequate equipment and other necessities. The teacher was Mr. W.H. Fry, and he earned sixty-five dollars per month.

In 1914, the Hot Springs County Boundary Board set the boundaries for School District #13. The next year, the boxcar was replaced by a gypblock building, which included a teacherage. Later, an additional room was added.

During the next several years, two teachers taught grades kindergarten through eighth; but in 1935, the student population decreased so drastically that only one teacher was deemed necessary. However, this decrease in students did not mean that the academics and athletic successes suffered in any way. For many years, Kirby continued to be "standard in its qualifications and it has a record that is most enviable in scholastic and athletic contests made by pupils in both their grade and high school careers."

The teachers who worked in Kirby during the earlier years included Frances Bennett, Elmer R. Flager, Lucile Beaty, Norma Miller, Ada F. Horr, Rosemary Gary, Lois MacCallum, Mildred Devoe, Sylvia J. Hammond, Wanja M. Birkey, Kathleen Talmage and Lavena D. Kagis. Later, Amy Hillier and Ella Johnston joined the staff.[300]

ORGANIZATIONS

In the mid-1920s, the ladies of Kirby decided to form a service group of their own, so the Community Club of Kirby was organized on January 29, 1925, with Mrs. Horr hosting. The following motions were passed: ten cents dues should be collected at each meeting; a lunch consisting of three "things" should be served at each meeting, with two members acting as hostess; meetings would be held on the second and fourth Thursday of each month; everyone in Kirby should be given the chance to join before the next meeting, so that "a set membership might be arranged at that time." Those present were Wallie Wright, Mildred Ellis, Adah Horr, Mae McManis, Florence Jenkins, Margaret Dusapin, Mabel Kruse, Alice Zube and Kathryn Graham. Several other ladies joined in the future months. It was later decided to become affiliated with the Federation of Women's Clubs, and a constitution and bylaws were drawn up. Members were asked to be declared as "active" or "associate" in status, and associate members could become active members whenever they wished. Offices had to be held by active members.[301]

Many civic activities were undertaken and sponsored through the years 1925 through 1928. The club raised funds to build and maintain a tennis court for the town. A bake sale was held, and twelve dollars was "realized." Two dollars were donated to the Boy Scout troop for work they did on the court. Weeds were cleared and posts set on the court property. The material for the court was hauled from Crosby, and a fence was erected around it. The members also investigated "conditions favorable to making a skating rink/pond north of the school." They were asked to subscribe to *McCall's* magazine, with fifty cents of each subscription to be turned over to the skating rink fund. Mrs. Dusapin offered to "stand for the expenses of lighting the rink, with the exception of globe and juice." The project was abandoned, as water was not available to flood the area to make the pond.

The club also provided Christmas gifts for the town's children, school age and younger. Members sewed garments for children of families in need. "Standardization" of the Kirby School was a major concern of the town, and the club was heavily involved in the discussions and processes needed to secure the necessary funding, equipment and materials to remodel the school building itself. The club learned how to sew and make carpet-rag rugs to be sold at a bazaar benefiting the school fund. Flour sacks from the local bakery were cut and made into dish towels. The members also purchased eighteen chairs for the elementary room of the school. A club member and her husband purchased a dictionary for the school, and the club purchased a stand for the new book from the Thermopolis High School. The club purchased as much "black board" as funds allowed for the school (eighteen feet long, three feet wide, for a total of $26.46); funds for molding and chalk troughs were also approved. The members also contributed to the fund to purchase a school bell. Two extra lots were needed to provide space for an adequate school playground. The club voted to pay Mr. Nelson of Kirby no more than $200.00 for the lots.[302]

LATER BUSINESSES

During the Prohibition years (1920 to 1933), government agents or "revenuers" made unannounced raids on establishments suspected of selling illegal alcohol. Almost always, the bottles made it outside and safe from prying eyes. Additionally, the moonshine was kept in a pitcher on the bar. When the "wrong fellow" came in, the moonshine was just poured down the sink. There were many ways to keep the hooch flowing. Sig Nelson related that he made many secret nighttime deliveries of wine grapes to Gebo, allowing homemade wine to continue to be secretly produced.[303]

Oil had been discovered in what would become western Hot Springs County, and it was easier to haul heavy equipment to the oilfields in Grass Creek and Hamilton Dome from Kirby than from Thermopolis. So the equipment was delivered to Kirby and dispatched from there.[304]

Earl Johnson's team hauls casing to the Grass Creek oilfields from north of Kirby.

A cute baby was possibly used by the Graham Dairy to advertise in 1936. *Ann Graham Robinson.*

Coal had been mined in Grass Creek for many years. While the tipple was still in operation in Kirby, coal was actually trucked to Kirby from Grass Creek and shipped from there.[305]

An article announced that Lucy Stevens of Kirby purchased two lots next to the Fletcher Garage in order to build an "up-to-date" ladies' fashion store. The store was scheduled to open as soon as the building was completed. It was expected to "fill a want that has been felt for some time by the ladies of Kirby and vicinity."[306]

Harold Van Buskirk bought a meat market in Worland in 1921. He "ran a pickup around the country, a peddling wagon." One day a week, he drove to Kirby and Crosby. The next day, he "worked" Gebo. Harold remembered that all of the people "liked legs of veal." He and his employee drove up and down the streets, especially in Gebo. They just knocked on every door.[307]

Ann Graham Robinson shared a history of her family's dairy in Kirby:

> *Alric and Kathryn Graham purchased the Tom Dempsey homestead, which was located three-and-one-half miles south of Kirby, on the river. Alric was the general foreman for the Chicago, Burlington, & Quincy Railroad at Kirby. He was in charge of the rip (repair in place) track, a side track where locomotives and/or railroad cars could be placed for minor repairs. He also oversaw the roundhouse operations.*
>
> *In 1923, their son Vern Paul Graham contracted typhoid fever, developed meningitis, and died at age 8. The Grahams had purchased milk from neighbors, and Vern's ailment was traced to contaminated milk from the neighbors' cows.*

The Grahams bought a cow to provide dairy products for their family's use. This enabled them to control the quality of the dairy products consumed by their family and ensured the safety of those products. The first cow they purchased gave too much milk for them to consume, so they sold milk to neighbors. Their cow went dry, so they bought more milk cows, and ended up supplying milk to all of Kirby.

Other residents in the area showed an interest in purchasing dairy products from the Grahams, so in November, 1926, they officially went into business as Graham Dairy. The Grahams' sons, Alfred T. "Lefty" Graham and Roy Graham, delivered milk for them. They eventually supplied milk to the Kirby, Gebo, and Crosby communities. An advertisement calendar, featuring a very cute baby, was printed in 1936.

In 1937, the Grahams sold the ranch and moved to Thermopolis. They sold Graham Dairy to Levere Steward, owner of Steward Creamery in Thermopolis.[308]

THE MINES CLOSE

In early 1938, it was reported that the Owl Creek Coal Company had closed down permanently. The loss of five hundred jobs in Gebo caused huge losses in Kirby as well:[309]

From Kirby it is reported that the depot at that freight division point will be closed as soon as the material is shipped out of Gebo. Also some stores at Kirby are planning to go out of business, however this has not been confirmed and is only a report.

The mine closing was not unexpected as reports for the past few years have predicted this closing, but it came as a shock to the people of the county just the same. Five hundred men have been employed at the mines since last August or September. The mines have not operated much in the summer months, being pretty much a winter operation.

The closing of the mine leaves a few people almost destitute.

The closing of the mines will probably nearly finish the town of Kirby and will be a lock to Thermopolis and Worland. However these places will have to develop other industries, refineries, agriculture, the springs' tourist business etc. to take the place of lost Gebo trade. A couple of small reservoirs on Owl Creek, and possibly on Kirby creek with the irrigation of land near Kirby on both sides of the river would bring in new farms, stabilize those we have and help take the place of Gebo.[310]

A Japanese Balloon Bomb

Duke Bolich shared a story from World War II that directly involved Kirby and Gebo.[311] In early November 1944, during World War II, Japan launched a number of bomb-delivering balloons. They were meant to travel on the jet stream and eventually land in the United States, creating chaos, fear and destruction. Two days after D-Day, at about 6:30 p.m., several people in Gebo saw a balloon floating over the northwestern part of camp. It was lit by a flare and floated very low. Most of the camp ended up watching it drift by.

After going over Gebo, the balloon floated eastward and landed near the Big Horn River east of Kirby. An explosives expert from Gebo, Steve Czarny, had followed the balloon's trajectory. When he went to examine it, he saw the explosives attached to the rim. He cut all of the wires attached to the bombs and threw the bombs in the river. With the weight of the bombs gone, the balloon floated free and became entangled in the overhead power line that serviced Kirby. Because the power poles were next to the highway, many onlookers were able to see the unusual sight.

On the following Monday, Pete Cavalli, who was a freshman in high school and riding the school bus back to Gebo from Thermopolis, asked the bus driver if he would stop and let the students look at the balloon. It was approximately thirty feet in diameter, and part of it was lying on the ground. Several of the students took a "souvenir" of the balloon. They were very surprised to discover that the balloon was made of paper.

In the following days, Duke remembered that the Hot Springs County sheriff's department came to the high school in Thermopolis and talked to all of the classes. The sheriff was joined by a man who was "an investigator from the Casper Air Base." The students were warned not to get close to any object that might be carrying explosives and to call the sheriff's office immediately if they saw anything suspicious.

The students were also "reminded not to talk about it because there might be Japanese spies listening, and he did not want them to know that one of the bombs came as far as Kirby. 'Remember, loose lips sink ships,'" he said.

News outlets in the United States were all told not to share information about balloon bomb sightings or landings. In the end, the bomb's components were traced to a facility in a beach area of Honshu Island (the main island of Japan). The area was attacked by U.S. B-29 bombers, and the balloon bombs were a threat for only six months.[312]

According to *Time* magazine, approximately 9,300 of the balloon bombs were launched. Because of the success of the "news blackout" regarding sightings and landings, it is not known exactly how many of the balloons actually hit their mark. Dr. Ross Coen from the University of Washington believes the first Japanese balloon observed in flight was the one that landed near Kirby.[313]

Family Stories

Kelly Punteney's family moved to Kirby in 1954. He shared the following reminiscence.[314]

The population was about 101, according to the 1950 census. The Punteneys were a large family: the twins, Ronnie and Donnie (who were too young to be in school yet), Kelly in first grade, his sisters Kay and Jean in second and his brother "Jeep" in third.

His dad (John Pershing, after the general) had bought their house in Kirby from the Deitchlers, who owned the gas station where John worked. The house was next to the school. For Kirby, it was a "really nice" house. It had two stories and four bedrooms and was 1920s style. It was heated with propane and coal. It also had a garden. Mr. Punteney had gotten a twenty-year mortgage for this house that cost only $5,000, because their family was "pretty poor."

At school, Kelly was the only one his age. He had gone to kindergarten in Thermopolis the previous year and then moved after school was out. Kirby had first through eighth grades in a two-room school, and Mrs. Hammond taught all the grades. Kelly had really enjoyed the bigger Thermopolis school and numerous activities. He didn't believe he was getting a "real" education like in town. He wanted to live with his grandmother and go to school in Thermopolis.

The 1950s were a hard time for the town of Kirby. The big mines had shut down in Gebo and Crosby because the railroads, businesses and homes had transitioned to diesel and natural gas. There weren't many "extras" at school either. Kelly did get to go to Denver on the train in seventh grade to see the Ice Capades. It was the first time he and his sister Kay got to travel out of the state of Wyoming. Several of the kids from Kirby got to go.

Eventually, the school was remodeled. All new desks were brought in to replace the old wooden ones, and restrooms were installed in the long hallway that went clear to the back of the building. They got a second teacher and so divided the big class into first through fifth (south classroom) and sixth through eighth, utilizing the smaller empty north classroom. The second teacher was Mrs. Hillier. Kelly's teacher was Mrs. Birkiehe.

The kids had to get creative when it came to having fun. They always went down to the river (all the railroad "stuff" had been abandoned there, including the pump houses that pumped water up from the river). They also went to Camels Hump mountain to the west of town. Kelly played with the Harpers, Kirkenslagers, Raiceviches, DeLeons (they lived in one of the boxcar houses) and Pam and Sue Watkins.

There was a roundhouse at the Y, near the stockyards, north and west of the tracks. Herb Harper lived in the maintenance (railroad) house, between the tracks and the railroad boxcar houses. Basically, the water tower was in their front yard. He was responsible for maintaining the tracks.

The railroad spur that had come from Gebo also wasn't in use anymore. However, the stockyards were still usable, and stock growers from all over the region loaded them in Kirby to ship them out. The boxcar houses were still there, lived in by railroad workers' families. The water tower was still there. It leaked like a sieve, but it was still filled. The train depot (the big one) was still there. The steam engines still went through Kirby every once in a while, going to different "final resting places." It was a big deal when they came through.

There was a small tipple in Kirby. Trucks came down from Gebo, and the coal was loaded from the coal trucks into the open coal cars on the railroad tracks. There was a spur where those coal cars were parked.

Things in Kirby slowed down year after year. Buildings deteriorated and were torn down over the years. The Green Door supper club and bar in Gebo burned down and was rebuilt in Kirby. The Burlington Northern train depot was still in Kirby when the Punteneys left in 1963.

As he grew older, Kelly realized that there were many things about Kirby that he liked. His kindergarten friends were his high school friends, too. He became a landscape architect after developing his love of gardening in Kirby. He learned his school lessons with the kids in the grades ahead of him, and they all listened to the other students' lessons, sitting off to the side. He learned excellent study skills.

He went from a school of twenty-one in Kirby to a high school of four hundred in Thermopolis to four thousand at Casper College in Casper, Wyoming, to forty thousand at the University of Oregon. Every time he went to a new school, he was really scared. But he "had an excellent educational foundation from his years in Kirby, so he studied hard and got good grades all through his schooling."[315]

According to Shelley Raicevich Deromedi, the George Raicevich family came to Minnesota from Montenegro around 1908.[316] He went back to

Montenegro in 1912 to find a wife. He asked a girl he knew to marry him, but she refused. Her sister Petrana "Pearl" agreed to become his wife and came back to America with him. They married in Minnesota. All four of their children were born there: Milorad "Mich," Alexander, Martha and Ida. George and Pearl owned a store.

Pearl had serious allergy issues. She wanted to "feel better," so they agreed to go west. They settled in Gebo so George could work in the mines, and they arrived in Gebo in 1928. George became a "timber man" in Mine #1.

Pearl was a homemaker and fed and did laundry for bachelor miners. She and George raised chickens and goats. He built a smokehouse to smoke meats and a sauna to use for baths and showers. George also made a great deal of wine. He, like many of the men, said he bought wine grapes because "they were needed to make jelly for the kids." A friend of the family, Veda Sharp, recalled "secret compartments behind pictures" in her house in Gebo. They were excellent hiding places for liquor bottles between the studs in the walls. Schlivovitza, a plum brandy, was also made.

George worked in OCCC Mine #1 until it closed, and then he worked in the truck mines. He never missed a day of work. Ida had a son, Alex Johnson. She bought Ida's Inn in Kirby when Alex was twelve.

Mich graduated from Gebo High School around 1932. He attended the University of Wyoming for three years as a business major. He dropped out to come home to Gebo and take care of Pearl. He possibly worked in Mine #1 in the summer and also in the CCC for a year. When World War II started, he was drafted and served for four years.

In 1948, Mich was helping his father build a cabin in the Big Horn Mountains when he met Josephine "Jo" Wantulok. She was working at Meadowlark Lodge, and Pearl introduced them. They married in June 1949 and came back to Gebo to live with George and Pearl.

Mich and Jo then moved to Sheridan, and Mich worked for the Burlington Northern Railroad. Their oldest child, Michael, was born in Thermopolis in 1950. Their oldest daughter, Shelley, was born in Sheridan in 1952. They moved back to Gebo, and their daughter Patricia "Trish" was born there in 1954. Mich and Jo worked for Ida at the inn, and they moved to Kirby. Marta "Marty" was born in 1960.

In Kirby, the kids attended a two-room school. Shelley had three teachers in seven years: Ella Johnston for four years, Amy Hillier for one year and Ernest Clark for two years. Because she was such a high-achieving student, Shelley skipped seventh grade. She always helped with the younger kids.

The 1966 eighth-grade graduates of Kirby School are shown at their graduation ceremony. *Left to right*: Dave Snyder, Mr. Millard, Shelley Raicevich, Susan Goff, Mr. Clark and Dolly Murrain. *Shelley Raicevich Deromedi.*

Kirby School students are shown in 1963. *Front row, left to right*: Randy Eaves, unidentified, Richard Kirkenslager, Shelley Raicevich, Patricia Raicevich, Gary Anderson, Mike Nelson, Reggie Nelson and Steven Eaves; *second row*: Miss Ella Johnston, Michael Raicevich, Lorene Nelson, Cynthia Kirkenslager, Terry Anderson and Mrs. Amy D. Hillier. *Shelley Raicevich Deromedi.*

The students were constantly hearing each other's lessons, and there was a great deal of repetition. They did blackboard work in front of the other students often, memorized poetry and wrote many book reports. Shelley knew how to read before she started first grade because of learning from Mike. She felt the kids also "gained self-confidence from helping younger students with their lessons. The school was the center of the community, and we saw the importance of cooperation and neighborliness," she added.

The Kirby students did not go into Thermopolis for school until ninth grade. There, they joined the other "country kids" from Lucerne, Hamilton Dome and Grass Creek schools. They had all competed against one another at various contests, track meets and so on when they were younger. In high school, they were all new to the kids who had already been going to school in town. Because they all rode the bus, they had to ride back home at four o'clock. If they wanted to be involved in sports or other activities, they had to go home with someone from Kirby who worked in Thermopolis.[317]

ENTERTAINMENT AND FUN

Leah (Schneider) Piel came to Worland with her family around 1909 in freight cars. They brought their whole lives with them and rode with everything.[318]

As she got old enough to date, she and her boyfriend (and later husband), Sam, attended dances in Kirby:

> *But on Saturdays* [sic] *nights, we would go to town. This was our big night of dating and everything.*
>
> *Sam would meet me in town. I'd go to town with my brothers. There were three of them, Fred and Paul and Adam. I would ride into town with them because it was quite a ways out there for Sam to come get me. So I'd meet him in town.*
>
> *Saturday night after Saturday night, we went to Kirby. They had free dances there. Wow, did we have a time. Everybody and their dog went to Kirby on Saturday nights. There was a bar on the other side of the dance hall which we would not dare have gone in. If my dad would have found out I'd ever gone to a bar, I'da been barred, believe me.*
>
> *But we went and we had so much fun. The guys, Sam and his brothers, would probably get a dollar or a dollar and a half on Saturday nights for all their work that they did. This is what they had to spend on their girlfriends. So we went to Kirby.*

> *Once in a while, fights would break out because some of these older guys—well, Now Sam was 20 or 21, he was old enough to have a drink or so. But he wouldn't go have one since I couldn't have one. So they never go booze it up. But there were some guys that would go out there and they would get drunk and there would be fights out there in the street. The cops straightened them out pretty well. I always wanted to go watch these fights, but Sam wouldn't have it. He said, no, we're not getting involved.*[319]

According to Shelley, the whole town of Kirby always did things together.[320] One favorite activity was ice skating on Coal Draw, after walking the half mile from Kirby. The parents cooked hot dogs, and everyone could skate all the way to the river. They skated during the day and also at night.

Christmas was always a favorite time, and caroling was enjoyed. They ended up at Sig and Maude Nelson's home, where they enjoyed hot chocolate and cookies. Maude also played the organ.

Jo usually organized potluck suppers. They were held at the school, and everyone played cards and board games. She also collected donations for funerals and hospital expenses.

The kids enjoyed hiking together. They were gone all day and took a lunch with them. They were only ten or eleven years old, but their parents didn't worry about them being gone that long. Mike and Shelley's future husband, Jerry Deromedi, spent a lot of time hunting and trapping together when they got older. Mike also hunted with H. Rudy Kowlok.

KIRBY, STILL STRONG

Kirby was founded and built on sheer determination. Kirbyites have continued to demonstrate that determination throughout the town's history. From filling the jobs needed to bring the railroad south; to "staying put" after the big mines closed; to continuing to service the truck mines, oilfields and freight businesses; to serving as the location of a world-class distillery, those who call Kirby home have been willing to do "whatever needs to be done."

From the very beginning to today: the Kirby jail is seen with the Wyoming Whiskey Shop. The distillery opened to the public in 2012. Both offer glimpses into Kirby's story. *Photo by Levi Shinkle.*

A Last Reflection

The lost coal district of Gebo, Crosby and Kirby existed at a time when people were willing to take risks—often life-threatening risks—for a better life for themselves, their families, their descendants. Many of them came from foreign lands to escape hardships and terrible dangers. Others came from points across America, looking for the American dream. Those who were willing to go down under the ground, to free the coal coveted by so many, took the biggest risks and faced those risks together with their fellow miners. The citizens of Gebo, Crosby and Kirby exhibited pride, true friendship, understanding of hard work and the satisfaction that comes from it, tolerance and appreciation for other nationalities, the ability to recover from terrible pain and sorrow, real understanding of the term "community" and absolute appreciation for the new opportunities they were given.

Those of us who are descendants of those remarkably strong individuals salute them, thank them and hope that we can honor them through the lives we have chosen to live.

APPENDIX

SAMUEL W. GEBO AUTOBIOGRAPHY—PRIOR TO 1906

Samuel Gebo was born in Ogdensburg, New York, on March 12, 1862. His family, whose original name was Gibeault (zhee-BOH), immigrated to Quebec, Canada, from France and then to New York State. He had six brothers and five sisters.

He worked on farms from age ten to age thirteen, making two to six dollars per month. He then went to work for Mr. Pool, who ran a resort. Sam rowed boats for fishermen at the resort and continued to farm as well.

A physician at the resort told Sam about the coal mines in Streator, Illinois. The miners were paid two or three dollars for a twelve-hour shift, 6:00 a.m. to 6:00 p.m., with an hour off for lunch. At that time, the bedrock in that part of Illinois contained large amounts of coal.

Sam was big for his age, and he asked the doctor about getting a job mining coal. He was assured that a gentleman named Doam Armstrong, a friend of the doctor, would take Sam down in the mine with him.

Sam stayed at the resort until September 1875 to complete his work obligation. The doctor's son, who was Sam's age, stayed with him at the resort and then took him back to Illinois.

Sam arrived in Streator on September 7, 1875, and went to work in the mine on September 10. He told himself, "I struck my calling." And, "I had an understanding with myself that I would own a mine of my own someday." Mr. Doam taught Sam how to mine, drill lumber and how to handle powder. In a short time, Sam became very "handy" at the mine. Also, eventually, he did own a mine and worked there for three years.

Sam Gebo's (1)
Statement January 30 - 1938
I was born in Ogdensburg, New york
March 12, 1862. My parents were French
Canadian I had 6 brothers, three older + three
younger, 5 sisters two older + three younger
All I can remember of my young days was work
From ten years to thirteen I worked on farms
for as low as $2 a month and up to $6 a month
The man I worked for the last two years
Mr. Pool kept a summer resort on the St. Lawrence
River. It was said to be the finest fishing stream
in the world at that time. I rowed the fisher men
in the day time. the rest of the day and part of the
night worked on the farm, getting the cows up
and milking before breakfast + after dinner
at night

There was a Doctor with his family, at the
resort, from Streator, Illinois, and I rowed the
boat for him. He told me about the coal mines
at Streator and miners were making $2.00 or $3.00 a
12 hr hour day. six oclock to six oclock, one hour
for lunch. I was big for my age, weighed 160 lbs.
I asked if he could get me a job in the mines
he said yes that he had a friend by the name
of Oran Armstrong and he would take me in the
mine with him. I told him I would go on the 1st of
September when my time was up. They left on
the 20th of July but he left his son that was

(1)

The Northern Pacific Railroad planned to open a mine at Sims, North Dakota. Gebo went for one winter and then traveled to Timber Line, Montana. There, he learned the difference between the coal seams in Montana and those in Illinois. He was known to have excellent skills in framing mine timbers. He stayed for eight months.

Gebo and his next partner, Bill Slack, left for Portland, Oregon, and Goose Bay. Next, they followed the railroad to Seattle, Washington. They took jobs at the Renton, Washington mine and stayed there a year. The Northern Pacific Railroad opened a coal mine in Roslyn, Washington. The company employed switch locks to lay the tracks over the mountains. The two men stayed there until the fall of 1888.

Coal mines were opening up in the Red Lodge, Montana area, and ten-hour days were available to the miners. Gebo and Slack went along. Gebo worked at least seven shifts each week; some weeks he worked nine. (At some point during this time, the pronunciation of Sam's last name became "GEE (hard *g* like the word *get*)-boh." It was to be pronounced that way only in this area.)

Sam went to visit his parents in Frazen City, Minnesota, and there he met his wife, Bertha. They married on Thanksgiving Day 1889 and left to go back to Montana the next day. He worked in the mine, owned a ranch and ran a milk dairy.

Gebo's next venture was taking up a coal claim on Bear Creek, Montana, in 1892. He hauled coal to Red Ledge over the hill. He accomplished this by using 112 horses: twenty-five 4-horse teams and two 6-horse teams. Sam made a $20,800 profit that year and then sold out to the Northern Pacific Railroad.

Sam's next stop was Fromberg, Montana, where he started another coal mine. There was another mine located in the nearby "Gulch," which was called the Gebo mine. However, it was worked out and abandoned. Gebo eventually sold his mine to the First National Bank of Omaha, Nebraska.

Oil was discovered at Butcher Creek, twenty-five miles west of Red Lodge. Sam and banker Lou Cruse founded a drilling company, which they ran for three years. They did find oil; however, they could not stop the flow of underground water. Gebo sold out to Cruse, who continued to unsuccessfully try to stop the water flow.

A Mr. Daily of the Anaconda Copper Company asked Gebo to inspect the Diamond Coal Fields in Wyoming. Sam declared it a "great mine."

Daily was also interested in a tract of coal land in the province of Alberta, Canada. The tract was thought to contain coking coal, which was needed in the steel-making process. Daily asked that Gebo ship $20,000 worth of mining machinery to Alberta. However, it was discovered that the coal was on a fifty-degree pitch. Gebo sank a slope (shaft) five hundred feet. The coal was broken and not good coking coal. Gebo wasn't interested in trying to market the coal, because they had a lucrative contract with eastern mines for their coke.

Sam had a reputation for knowing the potential for coal property. He was asked to examine such a property in Arkansas for a Mr. Frank. Gebo's recommendation was to refuse the property, but somehow Frank was forced to pay $90,000. The entire transaction was a total loss for him.

Frank was still committed to making money with Sam, so he promised any amount of money if Sam would make a "good start" in a new venture. Sam told him about the unsold machinery, clean coal and a thirteen-foot coal seam one mile from the railroad in Canada, with Daily. Frank wrote a check for $5,000, and the machinery was finally moved. The Frank mine and the town of Frank were prospering.

In the midst of all these happenings, Gebo's health was suffering. He spent a month with Dr. Dugant, a neurologist. The doctor recommended that Sam receive treatment in Paris, France, from Dr. Meuier. The three months Sam spent in France did wonders for his health. Because he spoke fluent French, he made numerous friends and established lucrative business connections in the north of the country.

When Sam returned to the United States, he found his wife and children doing well. The mine was doing well also: 4,000 tons per day were being mined. But on April 29, 1903, approximately 120 million short tons of Turtle Mountain slid, burying the eastern edge of the town of Frank and killing ninety-three people. The mine suffered catastrophic damage on the surface: all the coke ovens, two hundred mine buildings, the tipple and the whole railroad yard were wiped out.

The interior of the mine had little damage. However, it took thirty days to clear away enough debris to enter the mine and six months to return to full production. It was possible to produce coal on a limited basis during the rebuilding. To celebrate the completion of rebuilding, Mr. and Mrs. Gebo took a six-month vacation to France.

In 1905, the Frank mine was sold for $5 million to several of Sam's business connections in northern France. Sam's company had also acquired a coal concession (contract) with the Bulgarian government and the Turkish government on the Black Sea.

The Gebos, Mr. Frank and Frank's business manager, Mr. Sprigg, made a trip to Europe. They returned to America in the middle of July, and Mrs. Gebo became desperately ill four days later. She never left her bed afterward.

NOTES

Introduction

1. Kelsey Kehoe, *What Is Coal?*, Wyoming State Geological Survey, 2023.
2. Dorothy Buchanan Milek, *Hot Springs: A Wyoming County History* (Basin, WY: Saddlebag Books, 1986), 247.
3. Ibid.
4. Ibid.; Ray Shaffer, interview, Thermopolis, Wyoming, October 6, 2023.
5. Milek, *Hot Springs*, 247.
6. Thomas A.L. Nicholas, "The Gebo Mine and Miners," Resource Section, *Casper Star-Tribune*, March 23, 1969.
7. Milek, *Hot Springs*, 249.
8. Iris K. Guynn, "The Social Interaction among the Residents of Gebo, Wyoming," comprehensive paper, Black Hills State College, 1971.
9. Bill Deromedi, interview, Thermopolis, Wyoming, July 25, 2023.
10. Science Direct, www.sciencedirect.com.
11. Deromedi, interview.
12. "The Most Dangerous Gases in Mining," Howden.com/en-gb/articles.
13. Deromedi, interview.
14. Ibid.

Chapter 1

15. Annie Kowlok Jones, personal archive, Hot Springs County Museum and Cultural Center.

16. Ibid.
17. Ibid.
18. Milek, *Hot Springs*, 247.
19. Jones personal archive.
20. John Ralph and Ray Shaffer, tour of Crosby, Wyoming, July 28, 2019.
21. Jones, personal archive.
22. Joe Zulevich story, Annie Kowlok Jones, personal archive, 1979, 1.
23. Jones, personal archive.
24. Ibid.
25. Zulevich story, Annie Kowlok Jones, personal archive, 1979, 1.
26. Jones, personal archive.
27. Ibid.; Zulevich story, Annie Kowlok Jones, personal archive, 1979, 1.
28. Ruth Barham Kitzerow story, Annie Kowlok Jones, personal archive, 10.
29. Ibid., 9.
30. Ibid., 12.
31. Ibid., 9.
32. Zulevich story, Annie Kowlok Jones, personal archive, 1979, 1.
33. Jones, personal archive.
34. Zulevich story, Annie Kowlok Jones, personal archive, 1979, 2.
35. Jones, personal archive.
36. Zulevich story, Annie Kowlok Jones, personal archive, 1979, 1.
37. Mrs. A.M. Giddings, "Crosby," *Thermopolis Independent*, Women's Clubs section, March 26, 1925.
38. Jones, personal archive.
39. Dorothy G. Milek, "Museum Musings," *Thermopolis Independent Record*, June 7, 1980, 4.
40. Jones, personal archive.
41. Ibid.
42. Ibid.
43. Ruth Barham Kitzerow story, Annie Kowlok Jones, personal archive, 2.
44. Ibid., 3.
45. Ibid., 4.
46. Jones, personal archive.
47. Giddings, "Crosby."
48. Kitzerow story, Annie Kowlok Jones, personal archive, 2.
49. Jones, personal archive; Ralph and Shaffer, tour of Crosby.
50. Jones, personal archive.
51. Kitzerow story, Annie Kowlok Jones, personal archive, 10.
52. Jones, personal archive.
53. Giddings, "Crosby."
54. Ibid.

55. Kitzerow story, Annie Kowlok Jones, personal archive, 7.
56. Marilyn Jones Revelle, interview with Lea Cavalli Schoenewald, Thermopolis, Wyoming, August 17, 2023.
57. Kitzerow story, Annie Kowlok Jones, personal archive, 11.
58. Ibid., 12.
59. Ibid.; Jones, personal archive; Karla O. Christensen, "Crosby: Not Just a Coal Camp," essay written for a History of Wyoming class, Bill Bragg, instructor. 4.
60. Kitzerow story, Annie Kowlok Jones, personal archive, 11.
61. Christensen, "Crosby," 5; Jones, personal archive, 4.
62. Kitzerow story, Annie Kowlok Jones, personal archive, 8.
63. Zulevich story, Annie Kowlok Jones, personal archive, 1.
64. Jones, personal archive.
65. Kitzerow story, Annie Kowlok Jones, personal archive, 8.
66. Jones, personal archive.
67. *Thermopolis Independent Record*, January 13, 1933.
68. Jones, personal archive.
69. Ralph and Shaffer, tour of Crosby.
70. Ibid.
71. Kitzerow story, Annie Kowlok Jones, personal archive, 13.
72. Revelle, interview.

Chapter 2

73. Milek, *Hot Springs*, 249.
74. Samuel W. Gebo, handwritten partial autobiography, Carbon County Historical Society and Museum, Red Lodge, Montana, 1.
75. Ibid., 6.
76. Nicholas, "Gebo Mine and Miners."
77. Gebo, handwritten partial autobiography, 4a.
78. Nicholas, "Gebo Mine and Miners."
79. "Many Arrests for Coal Lands Fraud: Wilberforce Sully, Chairman of American Malt Board, Among Those Under Indictment. Got Dummy Land Claims, Five Dollars Paid to Each Person Making Out an Affidavit—6,072 Acres Taken Out," *New York Times*, September 8, 1909.
80. Wyoming Tales and Trails, Coal Camps, Gebo, wyomingtalesandtrails.com/coal.html.
81. "Many Arrests for Coal Lands Fraud."
82. Ibid.
83. Ibid.
84. Ibid.

85. Nicholas, "Gebo Mine and Miners."
86. Guynn, "Social Interaction among the Residents of Gebo."
87. Nicholas, "Gebo Mine and Miners."
88. Ibid.
89. "Many Arrests for Coal Lands Fraud."
90. "Samuel Gebo," Wikipedia, en.wikipedia.org/wiki/Samuel_Gebo.
91. Wyoming Tales and Trails, Coal Camps.
92. Ibid.
93. Thomas A.L. Nicholas, letter to Iris K. Guynn, April 25, 1971, private collection.

Chapter 3

94. Rufus J. Ireland Jr. and Thomas A.L. Nicholas, personal letter, Amityville, New York, May 1, 1968, 4.
95. Nicholas, "Gebo Mine and Miners."
96. Ibid.
97. Rufus J. Ireland Jr. and Thomas A.L. Nicholas, personal letter, Amityville, New York, May 1, 1968, 1.
98. Mary Corcoran Hanssen, "Resident Recalls Memories of Gebo: Early Boom Town," *Independent Record*, June 21, 1979.
99. Nicholas, "Gebo Mine and Miners."
100. Ibid.
101. Guynn, "Social Interaction among the Residents of Gebo."
102. Ibid.
103. Janet Zupan Philp, interview with Lea Cavalli Schoenewald, July 20, 2023.
104. Deromedi, interview with Lea Cavalli Schoenewald, July 6, 2023.
105. Guynn, "Social Interaction among the Residents of Gebo."
106. Ibid.
107. Ibid.
108. Deromedi, interview with Lea Cavalli Schoenewald, July 6, 2023.
109. Guynn, "Social Interaction among the Residents of Gebo."
110. Johnstone and Woolman family archive, 1997.
111. Ibid.
112. Ibid.
113. Ibid.
114. Guynn, "Social Interaction among the Residents of Gebo."
115. Mary Gordon, personal recollection of interactions with Pietila family, 2023.

116. Guynn, "Social Interaction among the Residents of Gebo."
117. Ibid.
118. Ibid.
119. Donald "Duke" Bolich, interview with Lea Cavalli Schoenewald, August 9, 2023.
120. Ibid.
121. Guynn, "Social Interaction among the Residents of Gebo."
122. Ibid.
123. Ibid.
124. Bolich, interview.
125. Guynn, "Social Interaction among the Residents of Gebo."
126. Ibid.
127. Ibid.
128. Ibid.
129. Ibid.
130. "What's in a Name," *Thermopolis Independent Record*, circa 1971.
131. Ibid.
132. Guynn, "Social Interaction among the Residents of Gebo."
133. Ibid.
134. Ibid.
135. Ibid.
136. Ibid.
137. Ibid.
138. Ibid.
139. Ibid.
140. Ibid.
141. Gary Giannino, interview with Lea Cavalli Schoenewald, October 1, 2023.
142. "Notes about Photographs," 1976, 1997, Box 1 and Digital, folder Gebo Mines and Employees, Mileva Maravic papers, American Heritage Center, University of Wyoming.
143. Donald L. "Duke" Bolich, *Gebo: Living in a Coal Camp through the Eyes of a Young Boy, 1932–1943* (Leander, TX: Bookwriters, 2023).
144. Deromedi, interview.
145. Guynn, "Social Interaction among the Residents of Gebo."
146. Mileva Maravic, 6309, Box 1, Folder Gebo Town, Main Street.
147. Guynn, "Social Interaction among the Residents of Gebo."
148. Ibid.
149. Ibid.
150. Iris Carter Cavalli, interview with Lea Cavalli Schoenewald, June 27, 2023.

151. Guynn, "Social Interaction among the Residents of Gebo."
152. Ibid.
153. Ibid.
154. Ibid.
155. Ibid.
156. Gebo typescript document, undated, Box 1, Folder Gebo (typescript), Coll. #6309, Mileva Maravic papers, American Heritage Center, University of Wyoming.
157. Nicholas, "Gebo Mine and Miners."
158. Guynn, "Social Interaction among the Residents of Gebo."
159. Ibid.
160. Ibid.
161. "Writings: Gebo Wyoming," 1976–97, Box 1 and Digital, Folder Gebo (typescript), Coll. #6309, Mileva Maravic papers American Heritage Center, University of Wyoming.
162. Guynn, "Social Interaction among the Residents of Gebo."
163. Ibid.
164. Ibid.
165. Ibid.
166. Phil Roberts, WYOHistory.org, November 8, 2014.
167. Guynn, "Social Interaction among the Residents of Gebo."
168. Hot Springs County Museum and Cultural Center, picture archive descriptions.
169. Guynn, "Social Interaction among the Residents of Gebo."
170. Ibid.
171. Bolich, *Gebo: Living in a Coal Camp*.
172. Ibid.
173. Guynn, "Social Interaction among the Residents of Gebo."
174. Ray Shaffer, Hot Springs County historian, interview with Lea Cavalli Schoenewald, October 1, 2023.
175. "Writings: Gebo, Wyoming."
176. Nicholas, "Gebo Mine and Miners."
177. Bolich, *Gebo: Living in a Coal Camp*.
178. Guynn, "Social Interaction among the Residents of Gebo."
179. Lea Cavalli Schoenewald, personal recollection.
180. "Writings: Gebo, Wyoming."
181. Ibid.
182. Ibid.
183. Ibid.
184. Guynn, "Social Interaction among the Residents of Gebo."

185. Nicholas, "Gebo Mine and Miners."
186. Ibid.
187. Ibid.
188. Guynn, "Social Interaction among the Residents of Gebo."
189. Nicholas, "Gebo Mine and Miners."
190. Ibid.
191. Ibid.
192. Ibid.
193. Ibid.
194. "Notes about Photographs," 1976, 1997, Box 1 and Digital, folder, Coll. Mileva Maravic papers, American Heritage Center, University of Wyoming.
195. Nicholas, "Gebo Mine and Miners."
196. Ibid.
197. Wyoming Tails and Trails, "Big Horn County," www.wyomingtalesandtrails.com/thermopolis3.html.
198. Lea Cavalli Schoenewald, personal knowledge.
199. Ibid.
200. Guynn, "Social Interaction among the Residents of Gebo."
201. "What's in a Name," *Thermopolis Independent Record*.
202. Bolich, *Gebo: Living in a Coal Camp*.
203. Ibid.
204. Milek, *Hot Springs*, 206.
205. Guynn, "Social Interaction among the Residents of Gebo."
206. Ibid.
207. Ibid.
208. Ibid.
209. Bolich, *Gebo: Living in a Coal Camp*.
210. Ibid.
211. Jerry Deromedi, interview with Thomas Schoenewald, June 16, 2022.
212. Bolich, *Gebo: Living in a Coal Camp*.
213. Guynn, "Social Interaction among the Residents of Gebo."
214. Ibid.
215. Ibid.
216. Ibid.
217. Ibid.
218. Marilyn Revelle Jones, interview with Lea Cavalli Schoenewald, October 15, 2023.
219. Ibid.
220. Rufus J. Ireland Jr., letter to Thomas A.L. Nicholas, May 17, 1968.

221. "Owl Creek Coal Company Going Out of Business; Will Liquidate and Sell Off Assets: Five Hundred Men Thrown Out of Work: Must Hunt Other Occupations or Locations. Mine Definitely Will Not Be Reopened. Coal Supply Said Not to Justify a Large Mine. Numerous Truck Mines May Supply Local Demand," *Thermopolis Independent Record*, February 4, 1938.
222. Mileva Maravic, "More About Gebo, Wyoming," Box 1 and Digital, Folder Gebo (typescript), Coll. #6309, Mileva Maravic papers American Heritage Center, University of Wyoming; Deromedi, interview, July 6, 2023.
223. Bolich, *Gebo: Living in a Coal Camp.*
224. Ibid.
225. Ibid.
226. Maravic, "More About Gebo."
227. Bolich, *Gebo: Living in a Coal Camp.*
228. Lea Cavalli Schoenewald, personal recollection.
229. Deromedi, interview, July 6, 2023.
230. Janet Zupan Philp, interview with Lea Cavalli Schoenewald, July 22, 2023.
231. Leo "Shorty" Roncco Sr., "Shorty's Saga—Part I and II," *Western Byways*, 1982.
232. Nena Roncco James, interview with Lea Cavalli Schoenewald, August 13, 2023.
233. Roncco, "Shorty's Saga."
234. Janet Zupan Philp, interview with Lea Cavalli Schoenewald, July 22, 2023.
235. Deromedi, interview, July 6, 2023.
236. Philp, interview.
237. James, interview.
238. Philp, interview; Carol Trusheim Waters and Janet Zupan Philp, interview, September 22, 2023.
239. Philp, interview.
240. Ibid.
241. Ibid.
242. H. Rudy Kowlok, interview with Lea Cavalli Schoenewald, August 31, 2023.
243. Ibid.
244. Ibid.
245. Ibid.
246. Barbara Heron Workman, interview with Lea Cavalli Schoenewald, October 12, 2023.

247. Ibid.
248. Ibid.
249. Ibid.
250. Ibid.
251. Ibid.
252. Ibid.
253. Bolich, *Gebo: Living in a Coal Camp.*
254. Patsy Urist Dorman, interview with Lea Cavalli Schoenewald, October 12, 2022.
255. "What's in a Name," *Thermopolis Independent Record.*
256. George Talovich, manuscript, 1977, 1 and Digital Coll. #6309, Mileva Maravic papers, American Heritage Center, University of Wyoming.
257. Ibid.
258. Ibid.
259. Ibid.
260. Ibid.
261. Ibid.
262. Lea Cavalli Schoenewald, personal knowledge.
263. Talovich, manuscript.
264. "Blackwater Fire of 1937," Wikipedia, en.Wikipedia.org/wiki/Blackwater_Fire_of_1937.
265. Vernal Woolman, Jack Woolman and Pete Diemer, manuscript, 1998, 1 and Digital, Coll. #6309, Mileva Maravic papers, American Heritage Center, University of Wyoming.
266. Milek, *Hot Springs*, 202.
267. Miravic, Gebo typescript document, undated, Box 1, Folder Gebo (typescript), Coll. #6309, Mileva Maravic papers, American Heritage Center, University of Wyoming.

Chapter 4

268. Milek, *Hot Springs*, 202.
269. "Townsite Suit," *Wyoming Tribune*, September 11, 1907.
270. Milek, *Hot Springs*, 255.
271. Hot Springs County Museum and Cultural Center, photo archive descriptions.
272. Milek, *Hot Springs*, 202.
273. Ibid.
274. Kirby, Wyoming, Town Council minutes, 1.

275. Ibid.
276. Ibid.
277. "Kirby Residents Relate Jailhouse Memories," *Thermopolis Independent Record*, September 25, 1980.
278. Ibid.
279. Kirby Town Council meeting minutes, 1916–51.
280. "Kirby Kernals," *Thermopolis Independent*, December 14, 1907.
281. Ruth Kennedy, oral history, Washakie County Museum, transcribed July 1988.
282. Milek, *Hot Springs*, 202.
283. Ann Graham Robinson, Graham family history.
284. "Agreement Has Been Reached: Will Mean Dawning of Wonderful Era of Progress in Basin," *Thermopolis Record*, February 15, 1908.
285. Ibid.
286. Ibid.
287. "Canyon Means Much to State: The Open Door to Commerce: 35-Foot Dam Will Supply Sufficient Power for Next 10 Years," *Thermopolis Record*, February 15, 1908.
288. Ibid.
289. Ibid.
290. Hot Springs County Museum and Cultural Center, photo archive descriptions.
291. "The C.B.&Q.R.R.," *Thermopolis Record*, August 31, 1922.
292. Ibid.
293. Lee Whiteley, *The Yellowstone Highway: Denver to the Park, Past and Present* (Denver, CO: Johnson Printing, 2001), vii.
294. Ibid.
295. Ibid.
296. Ibid., 122, 123.
297. Ibid., 128.
298. "Andrew Wagner Killed in Mine Accident Friday," *Thermopolis Independent Record*, July 5, 1929.
299. Dolly (Bury) Daniels, interview with Lea Cavalli Schoenewald, October 4, 2023.
300. "Historical Data of Hot Springs County Schools," *Thermopolis Journal*, D.A.R. Edition, July 15, 1938.
301. Community Women's Club of Kirby, monthly meeting minutes, 1925–27, Hot Springs County Historical Museum and Cultural Center, ARC.63.85.022A-C.
302. Ibid.

303. Milek, *Hot Springs*, 202.
304. Ibid.
305. Ray Shaffer, interview with Lea Cavalli Schoenewald, October 30, 2023.
306. "Ladies Fashion Shop for Kirby," *Thermopolis Record*, November 25, 1920.
307. Harold and Mabele Van Buskirk, oral history, Washakie County Museum, July 15, 1976.
308. Ann Robinson Graham, Graham family history,
309. "Owl Creek Coal Company Going Out of Business; Will Liquidate and Sell Off Assets," *Thermopolis Independent Record*, February 4, 1938.
310. Ibid.
311. Bolich, *Gebo: Living in a Coal Camp*.
312. Ibid.
313. Melissa August, "The Forgotten History of the Japanese Balloon Bomb That Killed Americans in World War II," time.com, May 5, 2023.
314. Kelly Punteney, interview with Lea Cavalli Schoenewald, June 18, 2023.
315. Ibid.
316. Shelley Raicevich Deromedi, interview with Lea Cavalli Schoenewald, July 1, 2023.
317. Ibid.
318. Leah (Schneider) Piel, oral history, Washakie County Museum, February 5, 1994.
319. Ibid.
320. Shelley Raicevich Deromedi, interview.

ABOUT THE AUTHOR

Researching and writing about Wyoming are two of Lea Cavalli Schoenewald's favorite endeavors. She previously wrote about a boarding school for girls in Laramie, Wyoming, in *Polished Corners of the Temple Wall*, for which she received awards from Wyoming Writers, Inc. and the Albany County Historical Society in 2012. She also co-authored an updated *History of Hot Springs State Park* in Thermopolis, Wyoming, with her husband, Thomas, in 2021. Schoenewald has also written a number of newspaper articles for one of the largest newspapers in Wyoming. She contributed to the *Casper Star-Tribune*'s *They Served with Honor: Vietnam*, having the honor of interviewing and documenting the stories of several Wyoming veterans in 2017. She has presented programs on the Ivinson Hall for Girls and Hot Springs State Park to numerous county historical societies and museums throughout Wyoming. Her favorite presentation was relating how her family came to live and mine coal in Gebo. She also taught a variety of subjects in four Wyoming school districts for thirty-five years.